Microbiology Made Easy

Jon Adams

Copyright © 2024 Jonathan Adams

All rights reserved.

ISBN: 9798322235873

CONTENTS

1 Tiny Tenants The Basics of Microbial Life .... Pg 6

2 Body Invaders Viruses and How They Operate .... Pg 19

3 Microbial Metropolis Understanding Bacterial Growth .... Pg 31

4 Fight or Flight The Immune Response to Microbes .... Pg 41

5 Invisible Engineers The Environmental Impact of Microbes ..... Pg 55

6 Microbial Harmony Symbiosis in the Microbiome .... Pg 66

7 Fermentation and Biotechnology Microbes at Work .... Pg 75

8 Outsmarting Microbes The Fight Against Disease .... Pg 87

INTRODUCTION

Welcome to 'Microbiology Made Easy', a journey through the wondrous and microscopic world that teems with life around and within us—a realm often left unexplored due to its sheer complexity. This book is crafted for the curious minds who have gazed upon the complexities of microbiology and hesitated at the threshold, daunted by its depth and breadth. Here, we unfold the tapestry of microbial life, from A-Z, with a narrative richness that paints a vivid landscape of this intricate science.

As your guide through the microscopic mazes, we invite you to view the microbes with new eyes. Rather than confront a barrage of intimidating terms, you'll encounter relatable analogies that bridge the gulf between abstract concepts and daily experiences. The life of a bacterium isn't so alien when likened to the bustle of a city, or the action of a virus compared to a stealthy spy infiltrating a secure fortress. These deep analogies are the touchstone of our exploration, bringing to life the drama and dynamism of the microbial world in a way that's both engaging and enlightening.

This book will sate the appetite of those yearning to decipher the enigmas of microbes without oversimplification, offering a clear, yet rigorous, expedition into topics like cellular machinery, genetic blueprints, and the ecosystems of the unseen. We'll delve into how these minute creatures affect everything from our health to the environment and the future of biotechnology.

Through 'Microbiology Made Easy', you embark on an adventure that promises not just facts and figures, but stories and insights. You'll witness the battles waged against pathogens, understand the symbiotic dances among organisms, and appreciate the masterful manipulation of genetic codes that underpin advances in medicine and industry.

Whether you're a student aiming to demystify your course material, a professional seeking a refresher, or a layperson with a spark of interest, this book is your passport to the wonders of the microscopic. Prepare to transform your perspective as we uncover the minutiae of microbiology, articulated with the kind of clarity that illuminates, the detail that fascinates, and the analogies that resonate. Welcome to a world where the tiny is tremendous, and the unseen is unforgettable—welcome to 'Microbiology Made Easy'.

TINY TENANTS THE BASICS OF MICROBIAL LIFE

Start by imagining a hidden universe that exists right under your nose. 'Tiny Tenants: The Basics of Microbial Life' is your personal guide to this universe, offering a glance at the incredibly small yet impactful organisms that live unseen all around us. Picture, for a moment, a world that is vital to your own, with inhabitants so tiny they could dance on the head of a pin. These minuscule marvels are microbes, and they play an enormous role in every part of life on Earth.

Think of a single bacterium as if it were a tiny factory. Inside, a series of processes take place that are as efficient and complex as any assembly line known to humans. Proteins are manufactured—think of these as specialized tools that allow the bacterium to function, grow, and reproduce within moments. All this activity, normally invisible to the eye, happens within each microscopic tenant, sustaining not just their own tiny lives but affecting the vast network of existence, including yours.

To understand these processes, let's use a familiar scenario. Envision a standard kitchen blender representing a microbial cell's structure. The lid on a blender keeps ingredients contained just as a microbial cell membrane holds in all the necessary components for life. Now, the setting or speed at which the blender is operated mimics the cell's metabolic activities—mixing nutrients at just the right pace to create the perfect blend, much like how a cell metabolizes nutrients to generate energy.

Microbes, though unseen, touch every aspect of our world, from the air you breathe to the soil in which plants grow. They're farmers, janitors, and builders in their natural habitats, playing roles that, if halted, would see our familiar world grind to an unexpected stop. They break down waste products into simpler forms that can be reused by plants—think of them as nature's recycling team.

Yet, these tiny tenants also have a dark side. Some can wreak havoc, causing diseases that range from the mild to the devastating. However, in the intricate dance of the microbial ecosystem, there's a delicate balance, with helpful microbes often keeping the harmful ones in check.

Now, let's layer on a bit more detail. Within this ballet of biological activity, microbes talk to each other using chemical signals—a system of communication as nuanced as any human language. These signals can dictate when it's time to proliferate, change behavior, or defend against a threat. Imagine if every light flickering in the skyline of a city sent a specific message to its observers—that's not unlike how these signals operate within microbial communities.

From the essential to the potentially perilous, microbes demonstrate an array of capabilities. As you dive further into 'Tiny Tenants', each term will be unpacked and each process laid bare, translated from the esoteric to the everyday. While the world of microbes is complex, the story of their existence is interwoven with ours, and understanding them is not just fascinating—it's fundamental to understanding the larger narrative of life on Earth.

Imagine you're shrunk down and whisked away into the bustling domain of 'Microbe City', a place of wonder hidden within a drop of water. It's an intricate metropolis where each tiny inhabitant has a job that's vital to the city's survival, much like the characters of a well-choreographed play. Now, visualize bacteria not as strange science terms but as the city's cleanup crew, tirelessly whisking away debris off the streets. Think of algae like miniature solar panels, dotting rooftops and churning sunlight into usable power, keeping the lights on and the engines humming in this microscopic community.

As you stroll down the metropolises' nano-avenues, each microbial denizen you pass by is hard at work, similar to how bees in a hive know their role and contribution to the collective buzz. The yeast cells are like the bakers of Microbe City, puffing up doughy bread with their bubbly exhalations. Meanwhile, the fermenting microbes are akin to a brewery, crafting fine brews that not only taste good but also leaven bread and ripen cheese.

The magic of 'Microbe City' lies not just in the individual tasks of its residents, but in how these tasks interlace to create a rhythm and flow, akin to the gears inside a clock—seemingly insignificant alone but indispensable together. In the grand narrative of life, these small actors play lead parts, demonstrating that in the great tapestry of existence, every thread, no matter how thin, plays a unique role in holding the design together.

This urban sprawl of microscopic life is consistently vibrant, an invisible realm humming with the energy of countless unseen organisms. As you weave through this tale of 'Microbe City', each concept will feel less like a scientific principle and more like a chapter of a lively storybook, with every unfoldment further enriching the tapestry of this miniature but immensely significant world.

Here's the lowdown on the bustling factories inside our microbial friends, using some homely analogies to keep things light and digestible:

- **Energy Production: The Power Plants**
    - **Glycolysis:** Picture this as a conveyor belt in a factory, breaking down glucose, the sugar 'staff meals', into smaller energy packets.
    - This process is like taking a full meal and breaking it down into snacks to keep factory workers energized throughout their shift.
    - Its role in the ecosystem is like the municipal power grid, providing essential energy to keep all parts of the city functional.
    - **Citric Acid Cycle (Krebs Cycle):** Think of this as the central hub where the smaller energy packets get processed into useable 'city power'.
    - Each reaction here is a workstation on the factory floor, refining the product into something the whole city can use.
    - In the grander ecosystem, it's like the local energy company distributing power to neighborhoods (or other cells and microorganisms).

- **Building Blocks Assembly Line: Biosynthesis**
    - **Amino Acids & Proteins:** Envision these as the factory's product line, assembling the building blocks to create machinery (proteins).
    - Just like a factory producing car parts, bacteria use these to build everything from structures to enzymes.
    - These 'products' are the microbe's contribution to the economy of Microbe City, supplying the materials necessary for building and repairs.

- **Waste Management: The Recycling Center**
    - **Decomposition and Fermentation:** This is the microbial waste recycling process, turning by-products into reusable 'raw materials'.
    - Similar to a recycling plant converting waste into new plastic or paper,

microbes turn waste back into materials the city can use or sell off.
 - The resulting by-products (like ethanol in fermentation) can be found in human products such as bread or beer, proving their wide-reaching impact.

- **Defense Department: Immune Officers**
 - **Antibiotics Production:** Some microbes produce antibiotics, akin to a factory creating defense gadgets against corporate espionage.
 - These tiny defenders can inhibit other rival microbes, similar to a security system that keeps competitors out of the business loop.
 - In broader ecology, it's like Microbe City having its own police force that ensures the community stays safe and balanced.

- **Trade and Export: Nutrient Exchange**
 - **Mutualism and Symbiosis:** Imagine some microbes like merchants trading essential goods with plants or other organisms in exchange for shelter or resources.
 - Just as international trade agreements benefit countries, mutualism benefits both microbes and their hosts, showing the importance of trade to the city's economy.

Each of these industrial processes powers the unseen metropolis, "Microbe City," with a lively energy and robust economy that parallels human cities. Through this vivid depiction, the complexity of microbial metabolism becomes a tangible and enthralling saga of life at the microscopic level, reminding us that even the tiniest of organisms have economies just as complex and essential as our own.

Step inside the bacterial cell and witness an architectural masterpiece of nature. Begin with the cell wall: think of it as a robust outer shell of a fortress, sturdy and imposing, it maintains the cell's shape and wards off unwelcome intruders. It's the security system, ensuring the integrity of what lies within. The plasma membrane, sitting just inside the cell wall, is akin to the selective gates of an exclusive community, deciding who or what gets to enter and exit, diligently managing the flow of nutrients and waste.

Beneath these layers, find the cytoplasm, a gel-like substance that could be compared to the busy hallways of a bustling corporate building, where the ins and outs of daily life unfold. The ribosomes scattered in this space are much like manufacturing hubs, complex machinery churning out proteins, the essential building blocks that keep the cell functioning optimally.

Zoom in further, and there's the nucleoid, a library of blueprints in the form of DNA. This is where the cell's grand designs are stored, instructions that guide the creation of all that the cell is and will become. Similarly, the additional structures, such as plasmids, are like the innovation teams in startups, carrying ideas that help the cell adapt and thrive under various conditions.

These elements work in concert, just as the systems in a skyscraper work together to create a livable environment. Each part is vital, playing a specific role that keeps the bustling micro metropolis alive. Explaining a bacterial cell in terms of a building's infrastructure brings an intricate world within grasp, framing these microscopic wonders in the context of a human-made marvel. It's a reminder that both the seen and unseen worlds are composed of structures and processes that, while vastly different in scale, are remarkably similar in principle and purpose.

Imagine walking into an architect's office. The first step in constructing a building is drawing up the blueprints. In the bacterial cell, this task belongs to the nucleoid region, where DNA lives. Here, DNA acts like the original sketch, with all the necessary details to create something magnificent. But you can't build straight from the original sketch—you need copies that workers can use. This copying process is called transcription. DNA is unzipped, and a complementary strand of messenger RNA (mRNA) is crafted, capturing all the crucial instructions in a form that can be taken to the construction site, ribosomes.

Now the mRNA blueprint gets transported out of the nucleoid and heads over to the factory floor, known as the ribosomes. Ribosomes are the cell's own manufacturing hubs, where raw materials are turned into useful products. The ribosome binds to the mRNA, sliding along it, reading the coded messages like a contractor reviewing a blueprint line by line.

This is where the building truly starts. Every three letters on the mRNA, known as a codon, call for a specific building block, an amino acid. Transfer RNA (tRNA) is the delivery truck for these specific amino acids, each matching up with the appropriate codon on the mRNA. One by one, amino acids are joined together in a growing chain, forming a polypeptide. This process is called translation and can be thought of as the construction workers meticulously laying brick after brick according to the project plan—the developing protein.

But a chain of amino acids isn't ready to be put to use yet. Imagine it as a raw steel beam that needs to be folded and trimmed into the final shape of a girder. Similarly, polypeptides fold and undergo modifications, turning them into fully functional proteins ready to perform a variety of jobs—like the girders, wires, and pipes installed in a skyscraper.

Proteins do everything from building cell structures, speeding up chemical reactions, to acting as messengers. They are vital both to the tiny bacterial city and the larger world it exists within, much like how the components of a building support not just the structure itself, but also form an integral part of the city skyline.

This elegant process shows not only the sophistication of cellular machinery but also highlights the universal building blocks of life. Zooming in on protein synthesis offers a glimpse at the cityscape of life, with every structure, every worker, every action contributing to the skyline of existence. And just like that, the complexity of the proteins turning genes into traits is laid out brick by brick, within reach and fully connected to the grand scheme of life.

Imagine walking into the ultimate car dealership, one that stretches further than the eye can see, showcasing every type of vehicle imaginable. This place doesn't just sell cars—it's the whole world of microbiology on display, from petite smart cars to massive, rugged SUVs. Each microbe is like a vehicle, designed for a specific terrain and purpose. The small, nippy bacteria are the city-smart compacts, zipping around efficiently. Some are even decked out as delivery trucks, transporting nutrients across different parts of the body.

Then, you have the sleek, fast viruses that are like sports cars, hijacking your cellular machinery with the efficiency of a high-performance engine. Not all of them are up to no good; think of the beneficial ones as the emergency vehicles, offering therapies like gene treatments. Meanwhile, fungi can be seen as the sturdy utility vehicles, breaking down waste or sometimes causing a bit of trouble.

Out in the back, you might find the archaea, those rugged off-roaders built for extreme conditions, thriving where other microbes might falter. Each type has its own niche, its own unique set of capabilities and tasks, just as each model of car is suited to particular needs and environments.

Understanding these numerous organisms is all about appreciating their design and utility. It's like recognizing what makes a convertible perfect for sunny day cruises or why a minivan is the best choice for family trips. When you view microbes with this lens, the complex world of microbiology becomes as relatable and understandable as choosing the right car for the right job. It's a vibrant, bustling ecosystem where every detail has its place and reason—every microbe, a vehicle poised for action in the vast, ever-moving life on Earth.

Let's take a deeper look at the bustling world of bacteria, treating them like the specialized utility vehicles of Mother Nature's majestic project. Bacteria are the all-terrain vehicles, adeptly navigating various environments from hot springs to the human gut, all while performing tasks crucial for life's continuity.

In the grand scheme, think of them as the cleanup crew of our planet. They're on the front lines when it comes to decomposition, breaking down the leaves in your yard into simpler substances just like recycling plants turn our waste into reusable materials. This not only keeps ecosystems clean but returns vital nutrients to the soil, helping maintain the natural balance.

Now picture a construction site where a utility vehicle delivers nitrogen—a critical component required to build proteins and DNA—to plants. Certain industrious bacteria pull off this feat in nature. They fix atmospheric nitrogen, which plants can't use, into a form that's like a ticket to an all-you-can-eat buffet for plant life. Gardens flourish because these microscopic workers supply the essential building blocks.

When it comes to energy, bacteria are the powerhouses, running on diverse fuel sources. Some thrive in oxygen-rich areas, running like cars on open highways, using a process known as aerobic respiration to generate their energy currency, ATP. Others can maneuver in an oxygen-free environment, engaging in fermentation—think of them operating like submarines running on battery power, undergoing chemical reactions to keep the energy flowing.

And then there's the role of producers and decomposers. Like mini-factories, bacteria turn sunlight, chemicals, or plant matter into new organic compounds, driving the base of many food webs. As decomposers, old organisms are the raw materials, which they dismantle and recycle into fundamental nutrients, so nothing in nature's economy is wasted.

These tiny microbes, much like the engines in our cars or the furnaces in our homes, churn silently, powering life's cycles. Each bacterium is a gear in a vast, unseen machinery that keeps the Earth healthy and habitable. As you sip on your coffee, consider the symphony of bacterial activity that supports the production of every bean, showcasing not just the might of the microbial world but its essential role in the familiar comforts of daily life.

Picture this: you've just stepped into the ecosystem of Microbe City, a bustling miniature metropolis, where each inhabitant has a vital role. It's a place where the unsung heroes, the nutrient-recycling street sweepers, tirelessly work behind the scenes. These are the bacteria that break down organic matter, transforming yesterday's refuse into tomorrow's resources, much like the cleanup crews clearing away the remnants of a festive parade.

Up high on the communication lines, you've got the message-delivering mail carriers - the signaling molecules that act as the town criers or the postmen of the cellular world. They dart from one cellular address to another, delivering urgent messages that coordinate activities like growth and defense. Imagine them zooming through the streets on Vespas, messages in hand, ensuring that every part of the city knows what's up and acts in harmony.

And then there are the architects and construction workers – the microbes crafting complex biostructures and the enzymes catalyzing chemical reactions, each ensuring that Microbe City's skyline is forever evolving, never the same day to day, just like our own urban landscapes.

Each microorganism plays its part, as essential to the success of Microbe City as our workers and professionals are to our own vibrant societies. It's like an ensemble cast where each character, no matter how small their role might seem, is key to the narrative. Their daily grind, often invisible, underpins the living, breathing rhythm of life itself, proving that the collective success of these diminutive denizens is as pivotal as that of any human workforce.

Step into Microbe Park, an ecosystem within the urban sprawl where life teems with activity, much like a corporate office abuzz with the daily hustle. Here, in this green oasis, the interactions of microorganisms are as complex and nuanced as the maneuvers in a company boardroom. Some microbes are like ambitious start-ups, finding their niche and flourishing with innovative survival strategies, such as forming symbiotic relationships with plants that provide them with the nutrients needed to grow.

Others are akin to seasoned corporations, dominating their space with well-established processes. Take, for instance, the nitrogen-fixing bacteria that act like key account managers, securing essential contracts that convert nitrogen from the air into a form that plants can absorb and utilize, akin to sealing a pivotal business deal that benefits all parties involved.

Competition is inherent in this park, as it is in business. Some microbial firms fight for market share, battling over resources or space in ways that mirror corporate rivalries. Yet, this ecosystem, like a savvy network of enterprises, also showcases the power of collaboration. Microbes engage in mutualistic relationships, exchanging services like waste management for sugars provided by the plants - a perfect example of quid pro quo that keeps the biological economy buzzing.

In this community, actions are interwoven, and outcomes are interconnected. The everyday operations of these microbial entities have cascading effects on the park's health, impacting everything from soil fertility to plant growth. This synergy ensures that Microbe Park remains a verdant, vibrant habitat, demonstrating a balance between the drive for individual success and the overarching need for a stable, functioning system – a delicate dance between self-interest and mutual benefit that encapsulates life itself.

Here's the breakdown of the nitrogen cycle, using terms you might hear during a coffee break chat about a thriving business project:

- **Nitrogen Fixation: The Client Acquisition**
    - Think of nitrogen-fixing bacteria as the go-getters of a business

development team.
- They take nitrogen gas from the atmosphere—something that plants can't use directly, much like an untapped market in business.
- These bacteria convert it into ammonium, a form plants can absorb, similar to signing a new client and bringing resources into the company.

- **Nitrification: The Production Process**
- Here's where the magic happens: ammonium is transformed into nitrates.
- Imagine a factory line taking a raw resource—ammonium—and refining it into a more usable product—nitrates, akin to a manufacturer preparing goods for the display shelf.
- Two types of bacteria are involved: first, the ammonium is converted to nitrites (like an initial product draft) and then to nitrates (the final, polished product).

- **Assimilation: The Sales Strategy**
- Think of plants in a garden as the sales team of this operation.
- They absorb the nitrates, which is like putting the finished product into action, ensuring the resource investment pays off.
- This is where the nutrients contribute to plant growth, akin to the sales team driving revenue with the product.

- **Denitrification: Client Feedback and Product Improvement**
- The cycle ends with the review process—denitrification.
- Here, different bacteria take nitrates and turn them back into atmospheric nitrogen, like a business acting on customer feedback to refine and better their services.
- This process closes the loop, essentially putting the unused resources back into circulation, preparing for another round of the business cycle.

Each step of this nitrogen cycle has a surprising parallel in the business world, painting a picture of microorganisms working tirelessly to support life's infrastructure from the ground up. It's not just a series of chemical reactions; it's a nature-driven economy where every action has a ripple effect—a meticulous network that mirrors our own complex world of business and industry, seasoned with a dash of natural innovation.

Venture into the inner sanctum of the human body, and you might find a scene not unlike a medieval fortress during a siege. Within these walls, a loyal legion of beneficial microbes stands guard, vigilant protectors of your health, much like the steadfast knights of old sworn to defend the realm. These tiny guardians uphold a fortress of wellbeing, fostering digestion, bolstering the immune system, and keeping the nefarious bacteria—the would-be saboteurs and stealthy spies—at bay.

These rogues, on the other hand, are the crafty intruders, lurking in the shadows and seeking a way past our natural ramparts. Cloaked in deception, they plot to slip undetected through the gates, eager to disrupt the peace and quiet of our corporeal city.

Yet, this is not just a tale of conflict. It speaks volumes about the delicate truce that reigns within us. The vigilant microbes are not merely repelling invaders; they are nurturing the land, interacting with our cells in mutually beneficial ways, much like farmers tending fields within the fortress walls, ensuring the land is rich and the populace fed.

It's this delicate balance—this constant juggling act between defense and nurture—that ensures the survival of the castle, our body. Neglect the guardians, and the rogues might find their chance. Care for your microscopic knights, and they'll care for the kingdom in return, creating a state of enduring health. It's an age-old story retold within us every day, a never-ending quest for harmony in the bustling metropolis that is the human body.

Picture the human body as a thriving empire, with a complex supply chain meticulously managed by probiotics, the tiny yet mighty workers within the gastro-intestinal fortifications. These beneficial microbes are like the loyal custodians of our digestive tract, ensuring that the delivery of nutrients to our cells is smooth and that the removal of waste is efficient. Here's how they work, step by logical step:

First, many of these friendly microbes aid digestion by breaking down fibers in food that we cannot digest on our own, similar to workers in a recycling plant processing materials to be reused. This not only aids in digestion but also generates vital nutrients, like B vitamins and short-chain fatty acids, contributing to overall health.

Second, imagine these probiotics as the body's own environmental health workers, regulating the pH balance in the gut like technicians managing a swimming pool's chlorine levels – making it less hospitable for harmful invaders, the pathogens that come knocking like burglars looking to break in.

As pathogens attempt to gain ground, the body's immune system, the elite royal guard, comes into play. Beneficial microbes act as scouts, detecting these threats early on. They communicate with the mucosal immune system – akin to watchtower guards signaling when an enemy is nearby – helping to prime the immune defense.

However, when rogue bacteria approach, they do so with a cunning arsenal. Some may secrete enzymes to break down the gut's protective mucus layer, like sappers undermining castle walls. The body's defense may respond with antimicrobial peptides or, in some cases, the deployment of antibodies – similar to pouring boiling oil over castle battlements.

In this eternal struggle, some bacteria may develop resistance to the body's antimicrobial tactics, much like invaders donning siege-proof armor or finding new strategies to breach the defenses. This arms race is as old as the human body itself, with each side consistently adapting and counter-adapting.

Through all this, the goal remains to maintain a flourishing empire where nutrients are abundant, waste is managed, and threats are adequately repelled. It's a highly sophisticated system, where much like a well-guarded fortress town, the presence of a healthy microbiome ensures that life continues without a hitch.

Understanding these intricate roles and processes of the microbiome in nurturing and protecting the body brings into focus the profound impact these microscopic organisms have – they're not just inhabitants of our body; they're active participants in maintaining the body's health, peace, and prosperity.

As the pages of 'Tiny Tenants: The Basics of Microbial Life' are gently closed, a moment of reflection unveils the staggering diversity and profound impact of microorganisms. These minute entities, formidable in their capacities, are now revealed as the unsung heroes and quiet architects of our world. From the depths of the oceans to the tips of the highest mountains, these lifeforms persist, influencing more than can be measured by their size. This journey through their world is not just about gaining knowledge; it's about fostering admiration for the intricate dance of life these microbes perform daily in every nook of our ecosystems.

Understanding them is not just an exercise in science; it's a widening of perspective, recognizing that within every drop of water, every breath of air, a complex narrative of survival, cooperation, and evolution is unfolding. They lay the foundations for life, break down the walls between death and rebirth, and connect every living being in an unbroken chain.

Each microbe, each process has been dissected and laid out with clarity—emphasizing how even the smallest component plays its part in the grand opera of existence. This understanding nudges every reader—whether peering through a beginner's lens or an expert's magnifying glass—to view the microbial world not as a distant concept but as a tangible, integral part of our own lives. Take this knowledge, this sense of wonder, and let it seep into the way you see the world, for within its smallest corners lies a universe brimming with life's tenacious whisper.

BODY INVADERS VIRUSES AND HOW THEY OPERATE

Step into the shadowy world of viruses, nature's own breed of secret agents. Just as spies weave through the crowds, unnoticed yet ever-present, so too do these microscopic operatives move among us. Each virus operates with a singular goal, carrying its encrypted genetic instructions across enemy lines into the unsuspecting cells of the body. One might think of a virus as a master of disguise, cloaking itself to bypass cellular defenses and usurp the cell's machinery for its own purposes, much like a covert operative turns an enemy's resources against them.

In this realm where biology meets espionage, understanding the virus's modus operandi is more than an academic pursuit—it's a journey into the very heart of survival and adaptation. As this chapter unfolds, the life cycle of a virus is laid bare with precision, from the initial stealthy incursion to the final, often-dramatic exit from the host cell. It's a tale of biological intrigue and sophistication, one where the stakes couldn't be higher: the health and well-being of entire populations.

Visualize each step as a critical phase in an undercover mission, where success and failure hang on the smallest details—the kind of narrative that doesn't just inform but captivates, driving home the impact of these viral agents with the weight they duly deserve. This chapter is your dossier, the briefing needed to understand these formidable agents of disease—and, ultimately, to appreciate the complex interplay between pathogen and host that defines so much of life on Earth.

Consider the influenza virus – the ever-adapting chameleon, changing its coat seasonally, always one step ahead of the scientist's predictions, much like a master of disguises evading capture. Then there's the rhinovirus, unassuming and common, the cause of the cold, which though it doesn't bring grandeur, certainly knows how to make an entrance and stick around, akin to the local rascal who knows every nook and cranny of your neighborhood.

On the other hand, picture the respiratory syncytial virus (RSV) as the silent creeper, which slips in unnoticed but for children and the elderly can stir up quite the storm, reminding one of the stealthy cat burglar who leaves chaos in the quiet of the night. Then, of course, there's the infamous HIV, the crafty infiltrator that disarms the body's defenses and slowly but surely takes over, much like a double agent that betrays from within.

And how could one overlook the new kid on the block, SARS-CoV-2, the cause of COVID-19? This virus has the world grappling with its sheer unpredictability and pervasive reach, much like a globe-trotting spy shrouded in mystery who slips through borders and stirs nations, leaving a trail of intrigue and urgency in its path.

Each of these viral characters plays its part in the ecosystem of human health, much like vibrant personalities in a bustling cityscape. Through understanding their unique 'personalities,' one can begin to appreciate the interplay of forces in the microscopic world that has a macroscopic impact, shaping our daily lives and our responses to these unseen yet ever-felt presences.

Here is the breakdown of the pathogenic processes of the influenza virus, spiced with analogies to bring the technicalities closer to home:

- **Rapid Mutation: The Virus's Disguise**
- Just as a seasoned spy might alter their appearance to remain undetected, the influenza virus frequently changes its surface proteins (antigens), thereby avoiding recognition by the immune system's 'special agents'.
- **Antigenic Drift:** Small changes in the virus's genes lead to slight alterations in its antigens, like a spy making subtle changes to their disguise to stay ahead of pursuers.
- **Antigenic Shift:** Occasionally, a significant change occurs, usually by combining genes from different virus strains, akin to a spy undergoing an extensive makeover to become unrecognizable.

- **Immune Evasion: The Stealth Operation**

- Influenza employs these mutations to slip past the immune response, much like an undercover agent slipping past security with fake ID.
- This constant changing of 'identity' makes creating an effective and long-lasting vaccine like trying to predict a spy's next disguise.

- **Infection Progression: The Mission Timeline**
  - **Initial Infiltration:** The virus enters the body and finds a host cell to hijack, reminiscent of a spy finding their target.
  - **Symptom Onset:** Within one to four days, symptoms appear—as if alarms are sounding due to the spy's actions.
  - **Viral Replication:** The virus forces the cell to produce viral components. Imagine a spy turning a company's printers to produce counterfeit currency.
  - **Cell Damage and Release:** The new viruses burst from the cell, destroying it—like a spy leaving a base in ruins after stealing what they came for.
  - **Immune Response:** The body's defense forces mobilize to combat the virus, similar to a rapid response team closing in on the chaotic aftermath of a spy's mission.
  - **Resolution or Complication:** The infection either gets cleared, or it escalates, potentially leading to serious health issues, as if the mission could either be foiled or lead to more significant security breaches.

The ways of the influenza virus echo the craft and guile of espionage. By understanding its covert operations and how they unfold within us, we become better equipped to intercept and counter its moves, safeguarding our health against this cunning opponent.

Think of viruses as tiny wanderers looking for a home where they can settle and multiply, but there's a catch—they can't build this home by themselves. They are not like bacteria that can live and grow on their own; viruses need to hijack the machinery of a living cell to replicate. Now, a virus itself is pretty straightforward in its design. It has a core of genetic material, which is like the brain containing all the necessary instructions. This core is wrapped in a protein shell that protects those precious instructions, much like a suit of armor.

When a virus finds a suitable host cell, it latches onto it and slips its genetic material inside. You can picture this like someone planting a tree in fertile ground. The virus's genetic material then uses the resources of the host cell—

similar to how a seed uses soil and water—to create more of itself. That's why viruses need a host to replicate: they simply don't have the tools to do it on their own. Understanding these basics gives us a clearer picture of what a virus is and how it operates on a fundamental level, setting the stage for deeper exploration into the ways we can defend against these microscopic invaders.

To truly grasp the stealthy nature of viruses, imagine witnessing the construction and deployment of a gadget from a spy blockbuster, but this all happens on an almost unimaginably small scale.

First, the virus finds a host cell—an operation target. Picture this as a spy finding the entrance to a secret facility. The virus uses particular proteins on its surface, which act like a set of lockpicks, to attach securely to receptors on the cell's surface. In our analogy, the virus's proteins are the spy gadgets designed specifically to gain entry into the facility.

Once attached, the virus enters the cell. Sometimes it merges its own membrane with the cell's, or it might trick the cell into swallowing it whole, much like our fictional spy executing a perfectly-timed slip past security doors.

Now inside, the virus unveils its genetic material, akin to the spy revealing the master plan. The cell, none the wiser, reads this genetic blueprint and sets to work, using its own machinery—think of factory workers unaware they are producing the components for their adversary's device—to replicate the virus's genetic material and construct new protein shells.

Gradually, piece by piece, new viral particles are assembled from these components. Just as a gadget comes together piece by piece on a factory line, each new virus is built and prepared for action.

Finally, these newly minted viral 'gadgets' emerge from the cell, often by causing the cell to burst open in a mass release, similar to the grand finale of a spy operation when the gadgets are used to complete the mission. Having left their cell factory in shambles, the new viruses are free to infect more cells, and the cycle begins anew.

Understanding this cycle not only paints an accurate picture of what a virus is and its replication journey but also underscores the necessity for precision in fighting viral infections—a high-stakes game of counterespionage waged in the microscopic world.

Imagine a virus as a crafty spy equipped with an array of gadgets designed for a single mission: infiltrating an enemy compound, represented by the human cell. To enter, the virus must first identify a specific entry point, just as a spy locates a secret door. These entry points are receptors on the cell's surface, and they require just the right key to unlock them. The virus's surface proteins act as these bespoke keys, expertly crafted to fit perfectly with the cell's receptors, allowing the virus to latch on tightly, like a glove to a hand.

This moment of connection is like the spy whispering the correct password, and the door swings open, granting them access. The virus then either merges with the cell's outer layer, slipping in like a shadow, or tricks the cell into swallowing it whole—much as a spy might employ a Trojan horse strategy to get inside.

Once past the cell's defenses, the virus reveals its true identity, unpacking its genetic blueprint and commandeering the cell's machinery to begin its covert operation. The cell, now a commandeered base, unwittingly starts replicating viral parts instead of its regular functions. Each step of the process is meticulously designed, as the virus effectively turns the cell into a factory producing more of the infiltrators. It's a sophisticated and delicate dance between identifying the target, breaching defenses, and taking control—all under the guise of normalcy until the viral objective is achieved and new agents are ready to deploy for the next mission.

Let's take a deeper look at how viruses master the art of cellular invasion, starting with a familiar handshake, the virus-receptor interaction. Imagine a virus as a master key fob that has been programmed to interact with a specific model of car—only the right model, or in our case, the right cell with the precise receptors, will respond to this key fob. The virus's surface proteins are the key fob, designed to fit effortlessly into the cell's receptor locks, forming a tight bond that establishes the initial secure contact.

Once the bond is formed, the virus undergoes a change in shape, like a spy using a disguise to slip past security undetected. This structural change can either fuse the virus directly with the cell's membrane, allowing it to slip into the cell, or trick the cell into engulfing it in an endocytic bubble—much like a spy hiding in a Trojan Horse to gain entry.

Inside the cell, the virus finds itself in a vesicle, which it must escape from as a spy would slip out of a poorly guarded room. The acidic environment inside the vesicle prompts the unpacking of the virus's genetic material, which is then released into the cell's cytoplasm.

Next, the viral RNA or DNA takes the cell's machinery hostage, repurposing it to replicate the virus's components. This hijacking can be likened to a spy taking over the command center and redirecting resources to produce counterfeit documents. The host's cellular machinery, now under viral control, begins creating new virus particles, and the cell transforms into a virus factory.

As the new viral particles assemble and mature, the process echoes the final stages of the spy's mission, culminating in the escape of new viruses from the hijacked base, ready to continue their covert operation elsewhere in the body.

This detailed espionage narrative not only makes it easier to grasp how viruses operate but also highlights the silent and efficient nature of their mission—to replicate and disseminate, perpetuating the cycle of infection. Through this story, the elegance, intricacy, and ruthlessness of the viral infiltrations come to life, underpinning the critical need for understanding which is crucial in combating these invisible adversaries.

When a virus replicates inside a host cell, think of it as a covert operation, where every step is precisely planned and executed. Once the virus has gained entry, the replication process begins. Consider the virus's genetic material as a set of plans for an intricate gadget—these plans must be duplicated meticulously. The virus commandeers the cell's own replication machinery to copy its genetic blueprint. This is akin to a spy using a photocopier in an enemy's office to make copies of secret documents.

Next, the virus takes over the cell's protein-making machinery. This is like bringing in specialized equipment to build the components of the gadget as per the copied plans. The viral messenger RNA tells the cell's ribosomes, the protein synthesizers, exactly what to produce. Just as a factory line might switch from making cars to bicycles, the cell stops its normal protein production and starts churning out viral proteins instead.

These new proteins then need to be assembled into new viruses, which is similar to putting together hundreds of intricate model kits. Each piece must fit together perfectly to form fully functional viruses that can go on to infect other cells. This assembly process takes place in the cell's cytoplasm, the fluid inside the cell, or at the cell's membrane.

Finally, once the new viruses are built, they leave the host cell. This final step can be like the newly built gadgets being packed and shipped out of the factory. Some viral particles push out through the cell's membrane and then break free without destroying the cell, while others cause the cell to burst open, releasing all the new viruses at once. This stage completes the virus's replication cycle, sending new viruses out to continue the operation and infiltrate more cells.

Understanding the viral replication process is like peering behind the curtain of a magic act. It reveals the nuances of the performance and emphasizes the skill of the performer—the virus—which is crucial for developing strategies to pull the plug on the show before it reaches its finale.

Let's take a closer look at the stealth-like process by which a virus infiltrates a host cell, reminiscent of a spy's mission to penetrate a high-security facility. The virus starts its mission in stealth mode, scouting for the right cell with specific molecules on its surface—these are the receptors, much like security checkpoints that grant access to the right individual.

When the virus spots its target cell, it uses unique proteins on its own surface, fashioned like custom-made keys, to lock onto these cell receptors. Once the virus finds its perfect fit, the entry process begins. This is the moment our spy has been waiting for—the gates are unlocked, access is granted.

This connection prompts the virus to undergo a structural change, adjusting its shape to strengthen the bond with the cell. It's like the spy changing disguises right at the doorstep, ensuring uninterrupted entry. Depending on the virus type, it may either fuse its own membrane with that of the cell, a seamless entry like slipping through a side door, or trigger the cell to engulf it in a vesicle—an endocytic bubble—much like hiding within a delivery package that's carried inside.

The virus, now inside the 'enemy' territory, sheds its stealthy cloak and releases its genetic material, the RNA or DNA, into the cell's cytoplasm. It's like the spy unveiling their hidden tools once safely within the compound's walls.

Next, the viral genetic material hijacks the cell's own protein-producing machinery. It commandeers the ribosomes, which typically produce proteins necessary for the cell's own use, to replicate the virus's genetic material and produce its proteins. This is akin to a spy infiltrating the command center and sending out orders to produce counterfeit documents.

The viral components are then rapidly assembled into new viruses. Imagine each viral component as a gadget component on an assembly line, carefully put together to create multiple copies of the original espionage tool.

The final phase of this covert operation is the release of newly assembled viruses from the host cell. Some exit by budding off, cloaked in a piece of the cell's membrane, while others burst out of the cell, leaving it destroyed. It's the culmination of the mission—replicates of our spy now infiltrate new terrain to continue their clandestine operations.

Understanding these mechanisms of attack and replication gives the insight needed to foil the viral espionage, developing antiviral strategies to interfere at every step of the virus's meticulous mission. This knowledge empowers us to bolster our cellular defenses and stay one step ahead in the ongoing battle against viral threats.

Picture our immune system as an elite security team in a high-stakes game of hide and seek with viruses, these crafty invaders with a knack for evasion. Like spies with an array of disguises at their disposal, viruses can change the proteins on their surface to avoid being recognized. It's similar to a chameleon altering its colors to blend seamlessly into its surroundings, escaping the watchful eyes of predators.

Some viruses take a different approach, akin to a seasoned actor slipping into various roles. They mimic the characteristics of healthy cells, effectively wearing a 'mask' that fools the body's security detail into thinking they're one of the good guys. The HIV virus, for example, is like a mole working on the inside, targeting the very cells meant to orchestrate the body's defense, weakening the immune response from within.

Then there are viruses that, like elusive ninjas, can lie dormant, hiding away in cells without triggering any alarm. This strategy of going undercover allows them to remain undetected by the immune system for long periods, only to re-emerge when defenses are low.

These cunning strategies highlight the viral world's ingenuity, keeping the immune system on its toes and scientists in the lab searching for ways to outsmart these microscopic escape artists. Understanding these tactics not only sparks our curiosity but is pivotal in the ongoing quest to design better defenses, ensuring our immune system is always one step ahead in this never-ending game of cat and mouse.

Let's take a deeper look at the art of disguise that viruses use to give the immune system the slip. Some viruses change their surface proteins through a process called antigenic variation, much like a chameleon changes its colors to match its surroundings, avoiding predators. These proteins on the virus's surface can switch to new forms that the immune system fails to recognize. It's as if the chameleon understands the vision of the predator and alters its colors in just the right way to become invisible.

Moving to molecular mimicry, some smart viruses can mimic the molecules found on the surface of healthy cells, like an undercover spy copying the mannerisms of a high-ranking official to move around a secured building freely. This mimicry fools the immune system into thinking the virus is a part of the body's natural landscape, letting the invader pass unchecked.

Now consider HIV, which is like a mole that infiltrates the headquarters of the immune defense: the T-cells. HIV integrates its genetic code into the host cell's DNA and can remain latent, quietly biding its time, similar to a mole that gathers intelligence and waits for the perfect moment to strike. During latency, the virus doesn't replicate much, which allows it to stay under the radar of the immune system. This retroviral latency is a form of immune suppression because it disables one of the central command cells of the immune response without triggering alarm bells.

Then there's the strategy used by herpes viruses, akin to ninjas who slip into a shadowy corner and go unnoticed. These viruses can enter a dormant state within their host cells, where they do nothing except exist as a silent genetic blueprint. They can lie in wait, undetected, for years, like ninjas waiting patiently in the shadows for the opportune time to act. When the conditions are just right, usually when the immune system is distracted or weakened, the viral ninjas emerge to strike.

Understanding these evasion tactics of viruses with the aid of analogies illuminates not just the what and how, but also the why—why these strategies are so effective and why it's so challenging to create vaccines and treatments. This insight is invaluable in our ongoing mission to outwit these microscopic adversaries and protect global health.

The recent global pandemic was caused by the novel coronavirus, SARS-CoV-2, the virus behind COVID-19. Initially identified in late 2019 in Wuhan, China, it spreads primarily through respiratory droplets when an infected person coughs or sneezes, much like throwing a stone in a pond causes ripples to travel outward. Contacts of infected individuals, touched surfaces, and even airborne particles in confined spaces became the vehicles for transmission, swiftly carrying the virus across borders and continents.

As the virus took hold, countries worldwide implemented measures to slow its march. These included lockdowns—akin to shutting down highways to stop the spread of a wildfire—quarantine protocols, and travel restrictions. Scientists and health experts launched an unprecedented counteroffensive, engaging in a race to understand the virus's behavior, vulnerabilities, and best defenses, such as wearing masks and social distancing, to keep communities safe.

Addressing the spread, medical systems were fortified, sometimes to the brink, teetering like a dam against a flood. The development and rollout of vaccines, which typically spans years, were compressed into months, illustrating the power of global cooperation and scientific advancement. Vaccination campaigns became the linchpin in this global effort, serving as the protective barrier between the virus and potential hosts.

The pandemic's spread and response showcased complexities of global health dynamics, human behavior's role in disease transmission, and the remarkable capabilities of modern medicine. Understanding this event provides invaluable lessons for managing future outbreaks and underscores the sheer resilience of societies in the face of microbial threats. This moment in history is not just a narrative of a virus but of civilization's response and adaptation in the face of a common unseen enemy.

Let's take a deeper look at the intricate web of strategies weaved together to combat the crafty SARS-CoV-2 virus. When it comes to public health measures, envision a multi-layered defense system in a medieval castle. Mask-wearing emerged as the sturdy gate, reducing the likelihood of enemy projectiles—virus-laden droplets—from entering the fortress. Picture social distancing as the open moat that keeps invaders at bay, the greater the distance, the less chance of an assault making it across.

Lockdowns were the strategy of retracting into the keep, sealing off entry points and limiting movement within the walls to prevent spies—silent carriers—from mingling with the inhabitants. Although disruptive, this tactic was designed to buy time, creating a lull in the battle to strategize and strengthen fortifications.

Turning to vaccine development, imagine if you will, blacksmiths forging powerful new weapons based on intelligence gathered from the enemy. The mRNA vaccines were like designing a training regimen that teaches the body's soldiers—immune cells—how to recognize and combat the virus without facing the real threat. This technology, previously on the fringes of vaccine research, stood out for its speed and adaptability, akin to a versatile new alloy that could be swiftly fashioned into formidable armor.

Clinical trials, meanwhile, were the rigorous drills and simulations ensuring the reliability of these new weapons. Conducted in phases, these trials meticulously gauged the efficacy and safety of vaccines across diverse populations, like a council of war carefully planning to launch a new defensive line.

Finally, the rapid scale of global vaccine distribution played out like an alliance of kingdoms sharing resources and knowledge to fortify their defenses. Despite logistical hurdles, akin to navigating treacherous terrain to deliver supplies, the rollout marked a pivotal moment in turning the tide of war against an invisible foe.

Understanding these diverse and sophisticated measures paints a vivid tableau of our world's collective endeavor to curb a modern-day plague. With the same camaraderie found at coffee tables, this discussion fosters a nuanced appreciation for the intricacies of public health and medical science, ultimately reflecting humanity's resilient spirit amid adversity.

In the narrative where viruses play the role of undercover agents, they exhibit a level of cunning that demands respect and vigilance. These biological entities, in their quest for survival, can breach our cell's security and replicate in secret, much like a spy who slips into a command base undetected. It's essential to understand their modus operandi; their methods of deception, evasion, and assault, to stay one step ahead. Unveiling the secrets of their stealth and subterfuge isn't merely an academic exercise—it's a critical endeavor that strengthens our defenses and shapes our strategies in public health.

Grasping the specifics of viral infection helps in designing precise countermeasures, from targeted therapies to vaccines that prepare our immune system for an impending attack. Much like decoding an enigmatic cipher, delving into the makeup and mechanism of a virus empowers us to pre-emptively dismantle its ability to harm. This understanding becomes a linchpin in the advancement of medical science, underpinning the fortification of global health security.

So, as one would arm themselves with knowledge to foil the plots of a cinematic villain, so too must our scientists, doctors, and everyday learners equip themselves with virological insights. This knowledge isn't just a shield to ward off current threats; it's a lantern illuminating the path to innovative breakthroughs, ensuring health and vitality remain within humanity's grasp. This enlightening dialogue opens the door to an empowered society where each individual grasps the magnitude of what's unseen, comprehending how vital the mastery of this narrative truly is for the wellness of all.

MICROBIAL METROPOLIS UNDERSTANDING BACTERIAL GROWTH

Picture a city skyline in the throes of construction, where once there was empty space, now towers a mesh of skyscrapers, roads, and parks, bustling with energy. In a similar sense, bacterial colonies rise from what was once barren terrain, a single microbe multiplying into millions. This growth unfolds when conditions are just right, like a metropolis blooming in a goldilocks zone perfect for expansion. Given moisture, nutrients, and a hospitable temperature, microbes thrive, forming colonies that mirror our urban sprawl. As you trace the outlines of these microbial cities, each blooming colony elucidates a complex story of life, replicating at a pace and scale that's astonishing, developing complex communities as diverse and dynamic as the inhabitants of New York or Tokyo. This spectacle of growth, both micro and macro, weaves a narrative of existence—one of resilience, expansion, and the relentless push towards life.

Imagine a solitary bacterial cell alighting on an oasis of nutrients—the perfect plot of land, if it were a homesteader in the old frontier. This single microscopic settler, armed with genetic blueprints and the simple machinery to multiply, wastes no time. In this land of plenty, it splits in two, then four, much like a pioneer's family growing as they build their life from the ground up. Soon, a small cluster of cells emerges, reminiscent of a circle of cabins in a clearing.

As long as the soil is rich with food—amino acids, sugars, and other life-giving molecules—the colony balloons, developing structures and specialized areas, akin to a homestead growing into a hamlet and eventually a town, with each cell playing its part in this tight-knit community. Waste products are managed, internal resources are shuttled to where they're needed, just as a settlement would manage its sanitation and trade goods.

These steps, each critical and precise, chart the course from solitude to society. A complex network of interactions, silent to the naked ear but as bustling and vital as any human hub, is born and blossoms within hours or days. This rapid transformation, from loner to legion, is life on the microscopic scale—silent, yet astoundingly swift, evident of the raw potential and power held in the smallest of life's forms.

Let's take a deeper look at binary fission, where a bacterial cell starts its journey of replication, much like a plot of land about to be split into two new properties. Visualize the bacterium's DNA as the original deed to the property. When conditions are ripe, the DNA replicates, like making an exact copy of the deed before the property division. Following this, the cell begins to elongate, not unlike a piece of land being measured and marked for separation.

As the bacterium stretches out, a partition starts to form in the center, known as the septum, similar to laying down a new fence line between the two plots. When the construction of the septum is complete, the cell pinches off to create two separate but genetically identical daughter cells, each inheriting a copy of the DNA—much like each new plot now has its own deed.

This efficient process allows bacterial colonies to grow swiftly in nutrient-rich environments, echoing the way a town expands rapidly during a housing boom. As the community grows, these bacterial cells, akin to townspeople, begin to differentiate to take on specialized roles, such as nutrient processing or waste management, ensuring the survival and prosperity of their microsociety.

Consider how a town manages waste and redistributes resources—this happens in a bacterial colony too. Cells expel waste through their walls and move resources around to avoid build-up and to maintain a healthy living environment, ensuring the community functions smoothly as a whole.

By exploring binary fission through the analogy of developing land and building a society, the details of this biological process come into clearer focus. It's not just an academic concept; it's a living, breathing event that mirrors the cooperative and dynamic structure of our own cities. Through this lens, the complex world of microbial replication becomes as tangible and relatable as the growth of our familiar urban landscapes.

Imagine a small town at the brink of a gold rush, where news of wealth has just hit the ears of eager prospectors. In the bacterial world, the exponential phase of growth is like this very moment, when the once-sleepy town burgeons into a bustling city. Each bacterial cell is a prospector, and the gold is the abundance of nutrients surrounding them. As they strike nutritional gold, these cells divide with unrestrained zeal, their numbers doubling in perfect harmony with the rhythm of ample resources.

This period is akin to a time-lapse of a skyline where buildings shoot up in rapid succession, roads widen, and populations swell. Infrastructures within this microbial metropolis expand and specialize. Like a town setting up various districts - residential here, commercial there - the bacterial colony develops complex zones to handle different functions.

As the colony grows, so does its vibrancy and complexity, a testament to the power of exponential progress. From each individual to the collective whole, the transformation is stark. A once quaint township transforms into a sprawling urban landscape within days or even hours, and in the petri dish of life, the microbial city knows no bounds. This phase of bacterial growth isn't just a scientific wonder; it's an embodiment of life's unyielding urge to flourish, painting a picture of unbridled potential that even the smallest organisms possess.

Here is the breakdown of the cellular processes during the exponential growth phase of bacteria, laid out like a town blossoming into a prosperous city:

- **The Alert for Expansion: Responding to Environmental Cues**
  - **Nutrient Availability:** Like news of a gold rush, an abundance of nutrients signals bacteria to start multiplying.
  - **Suitable Living Conditions:** Steady temperatures and comfortable pH levels act as the welcoming town square where growth can prosper.

- **Prepping for Growth: The Bacterial Cell Cycle**
  - **DNA Replication:** This is the drafting of the master blueprint for the future city; the cell duplicates its genetic material to share with its offspring.
  - **Cell Growth:** Much as a town expands its boundaries to accommodate more residents, the bacterial cell increases in size.
  - **Component Synthesis:** The cell manufactures vital building materials, akin to a city laying down utilities ahead of a population boom.

- **The Division of Labor: Cytoplasm Partitioning**
  - **Septum Formation:** A new wall forms in the cell's center, like dividing a metropolis into two manageable districts.
  - **Segregation of Cell Components:** Each newfound 'district' gets its fair share of the cellular apparatus, ensuring both can function independently.

- **The Birthing of New Cells: Binary Fission**
  - **Cell Wall Pinching:** Picture it as a ribbon-cutting ceremony, where the new partition is finalized.
  - **Daughter Cells Emergence:** Two separate yet identical bacteria emerge, ready to thrive, like twin cities born from the expansion drive.

- **Rapid Multiplication: Exponential Growth Explained**
  - **Resource-Driven Reproduction:** Bountiful nutrients and space allow for unfettered growth, as boundless resources would in a booming town.
  - **Genetic Optimization:** Bacteria are genetically equipped to utilize the nutrient 'wealth' efficiently, similar to a town with skilled prospectors ready to mine gold.

Like a once-quiet town erupting into a hub of opportunity, each bacterial cell replicates rapidly when conditions mimic the perfect storm of prosperity in a gold rush era. This phase is essential to understand, not just for the spectacle of growth it presents, but for the implications it has: in medicine, where it guides antibiotic use; in industry, as it shapes fermentation processes; and in environmental management, where it informs the balance of ecosystems. This session, reimagined within the context of a blooming city, is a cornerstone for grasping the potent, invisible forces that underpin the liveliness of both micro and macro worlds.

Picture a city that's hit the limits of its expansion – the land is developed, every apartment is occupied, and resources like water and power are maxed out. In the microscopic realm of bacteria, this is the stationary phase, where the population's exponential growth halts. Bacteria, much like city officials, must now get creative to sustain their society in the face of limited resources. Each bacterial cell, once thriving on the abundance, now faces competition for nutrients, space to grow, and begins to pump the brakes on replication.

This phase in bacterial life is a period of strategic adaptation where, much like a city under resource strain, new ways to thrive are sought out. Efficiency becomes key. The bacterial community has to manage waste more meticulously and allocate energy to survival rather than growth, ensuring that even amid scarcity, life endures.

In the broader ecosystem, this phase represents a balance, a crucial stabilization that echoes the need for sustainability in our own urban landscapes. Understand this, and the mechanics of population dynamics and the delicate dance with the environment become clearer, from the cell to the city.

When bacteria enter the stationary phase, it's as if a city has enacted strict conservation laws due to a sudden resource shortage. Here's a detailed look at how these microbial inhabitants manage to pull through:

First, as nutrients become scarce, the bacteria tighten their belts, slowing down the machinery that helps them grow and divide. In cities, this is much like initiating water restrictions or rolling blackouts to conserve water and electricity.

The bacteria's DNA, the blueprint for its survival, must be shielded from the stress that can cause damage. They start producing proteins that act like emergency response teams, repairing any harm and fortifying the DNA against further assaults. This can be likened to a city reinforcing its levees before a flood.

To conserve energy, bacteria reduce their metabolic rate—the speed at which they carry out life's processes. It's akin to a city turning off public lights or reducing public transportation schedules to save electricity.

Then there is the management of waste products, which in dense populations can build up and become toxic. Bacteria activate removal processes, similar to a city increasing its waste recycling efforts during times of crisis to prevent pollution.

Finally, some bacteria might even enter a state of dormancy, halting all activities as though they were battening down the hatches in anticipation of a prolonged resource drought—a mirror to a city in lockdown, waiting for a crisis to pass.

Throughout this phase, the bacteria's survival hinges on their ability to be resourceful and resilient, to quickly shift gears from a mindset of growth to one of sustainability. This biological resilience reflects the kind of crisis management and creative problem-solving that can keep a city afloat during tough times. Understanding the stationary phase highlights the sophistication underlying bacterial survival and the parallels to our own approaches to managing scarce resources.

Imagine a city facing an impending storm; officials roll out a robust disaster response plan, with everyone retreating to safe zones, vital records secured in vaults, and resources stashed away. In the bacterial realm, the formation of endospores is such a masterclass in crisis management. When the going gets tough—be it extreme heat, freezing cold, or chemical threats—some bacteria shift gears from their normal life-sustaining processes to lock down into an almost indestructible form, the endospore.

This process is akin to the construction of a high-tech bunker, built to withstand anything nature throws at it. First, the bacterium makes a copy of its DNA—the city blueprint—and encases it in a sturdy shell made of tough proteins, like a vault lined with steel. Then, it strips down, shedding unnecessary parts, much like a city conserving resources by turning off non-essential services.

What's left is a microscopic fortress that can survive conditions that would spell doom for other bacteria. It's not quite living, but it's not dead either—it's in stasis, waiting patiently for the signal that it's safe to emerge and begin life anew. Upon receiving this signal, much like a city sounding the all-clear, the endospore transforms back into a normal bacterium, ready to grow and multiply once more.

This remarkable strategy isn't just a marvel of the microbial world; it's a testament to the ingenuity of life. Much like our own urban centers preparing for the worst, bacteria with the ability to form endospores show us the art of survival, pulling through unscathed while the rest of the world spins into chaos.

Here is the breakdown of how a bacterium battens down the hatches to form an endospore, much like a city fortifies itself against an impending disaster:

- **The Triggering Alarm: Environmental Stress Signals**
  - Just as a city's emergency systems detect threats like approaching hurricanes, bacteria sense environmental stressors such as nutrient depletion or extreme temperature changes, signaling it's time to switch to survival mode.

- **The Blueprint Copy: DNA Replication**
  - **Core Encasement:** The bacterial DNA, which holds all the instructions for life, replicates and is safeguarded within a thick, durable coat resembling a city securing vital records within a fortified vault.
  - **Layered Protection:** Multiple tough layers surround the genetic material, much like a command center's walls shielding against the storm.

- **The Fortress Construction: Spore Coat and Cortex Formation**
  - **Spore Coat:** A hardy, armor-like shell forms around the bacterium, akin to a city barricading its windows against fierce winds.
  - **Cortex:** Beneath the coat lies a spongy layer called the cortex, functioning like shock absorbers designed to mitigate any tremors from the environment.

- **The Conservation Plan: Dehydration and Metabolism Shutdown**
  - **Water Removal:** The cell eliminates water, preserving itself nearly indefinitely, similar to a city rationing supplies during a siege.
  - **Metabolic Cessation:** The cell shuts down its normal activities, entering a dormant state like a city in lockdown, conserving power by turning off all non-essential services.

- **The Wake-Up Call: Germination**
  - Upon detecting favorable conditions, akin to a city's sensors picking up clear weather, the endospore reactivates.
  - The tough layers break down, and the bacterium absorbs water, revving back to life as a city might restart its engines, signaling the end of the crisis.

This magnificent adaptation is much more than an emergency protocol; it's a lifeline thrown across the abyss of time, allowing bacteria to emerge unscathed from periods that would otherwise mean certain doom. In understanding this phenomenon, the parallel capacities for resilience become clear, painting a picture of life's resourcefulness from the scale of single cells to sprawling cities.

Imagine a visionary mayor overseeing the growth of a city, meticulously planning each district to ensure a thriving, harmonious community—this is not unlike the process of creating probiotic yogurt. The dairy 'landscape' is first enriched with select nutrients, setting the stage much like an urban planner would prepare the grounds for construction. Into this environment, beneficial bacteria—like Lactobacillus and Bifidobacteria—are introduced as the city's first citizens, with a mission to transform milk into yogurt.

Under the careful control of temperature and conditions that mimic a mayor's diligent governance, these bacteria get to work. They ferment the lactose, the milk's natural sugar, producing lactic acid—a process akin to laying down the foundation for the city's infrastructure. This acid causes the milk to thicken and develop the tangy flavor characteristic of yogurt, much like the cultural essence of a town forms over time.

Quality checks along the way ensure that the bacterial population remains healthy and active, just as a city's services would monitor for the well-being of its inhabitants. The final product, with its creamy texture and beneficial properties, is a testament to the skilled orchestration of various elements in production, from ingredient selection to precise temperature control—resulting in not just a food item but a microcosm of microbial life ready to join and enrich the human gut's microbiome.

In every spoonful of probiotic yogurt, there is a story of controlled biological artistry, reflecting the delicacy with which we can coax life to work in our favor, much as a competent mayor who steers urban growth to benefit all residents. It bridges the gap between food science and city planning, illustrating the ingenuity behind fostering beneficial life, be it in a bowl or in a bustling metropolis.

Let's take a deeper look at the precision-crafted process that turns everyday milk into probiotic yogurt, akin to a finely tuned urban ecosystem flourishing under the careful eye of its planners.

First, we introduce the 'city founders,' beneficial bacteria such as Lactobacillus and Bifidobacteria, into the milk—a fluid landscape ripe for development. These bacterial settlers get straight to work, breaking down lactose, the milk's naturally occurring sugar, like industrious workers laying the groundwork for future structures. The lactose is their raw material, fermented and converted into lactic acid, which acts much like the mortar that cements a city together. This acidification process is key—it fortifies the yogurt, turning it from a liquid to a creamy semi-solid, mirroring the transformation of a landscape into urban sprawl.

As the milk's pH drops due to the increase in lactic acid, the environment becomes perfect for these beneficial bacteria to thrive, paralleling the careful zoning of a city to create an optimal living space. Their enzymes then get to work on the milk's proteins, particularly casein. The casein proteins, once free-floating like commuters in a city, begin to coagulate into a matrix, forming a solid framework—the city's skeleton.

With temperature control akin to a city's climate management and incubation times reflecting meticulous civic planning, the milk is steered through the stages of fermentation. These carefully monitored conditions ensure that the bacteria's activities are productive and sustained, mirroring how city managers would constantly gauge the efficiency of utility use or traffic flow.

The endpoint is a harmony of textures, flavors, and beneficial microbial life—a bustling metropolis of nourishment. It's not just food creation; it's the orchestration of living organisms creating a product that's more than the sum of its parts. In every cup of probiotic yogurt, there lies a complex dance of life sciences and culinary craft, a testament to the symbiotic relationship between humans and the microorganisms that support our well-being. It's a tale of transformation, made possible by the natural alchemy of fermentation—a process as beautiful and intricate as the evolution of cities from the ground up.

Consider the role of a city planner, tasked with shaping urban spaces to be functional, pleasant, and sustainable. Just as the planner carefully aligns roads, zones residential areas, and ensures that parks and public services enrich lives, so too must scientists and health professionals grapple with the intricate details of bacterial growth. The balance within our bodies, much like the balance within urban landscapes, relies on managing the growth of beneficial bacteria while curbing the spread of harmful ones.

Understanding these microbial dynamics is key to harnessing beneficial bacteria to our advantage. This could mean deploying probiotics to aid digestion, much like introducing green spaces to combat urban heat islands. It involves recognizing when harmful bacteria threaten to overwhelm the system, similar to how unchecked urban sprawl can put a strain on resources and quality of life.

At its core, managing bacterial growth for health is not dissimilar to urban planning. Both require a deep appreciation of complex systems, the vision to foster growth in a controlled, beneficial manner, and the foresight to prevent the detrimental effects of uncontrolled expansion. With the same precision a planner uses to craft a cityscape, experts scrutinize the microbial world, ensuring the delicate balance of our internal ecosystems supports a state of wellness, much as a well-planned city supports the thriving of its inhabitants.

FIGHT OR FLIGHT THE IMMUNE RESPONSE TO MICROBES

Step through the gates into the intricate realm of the immune system, an astonishing network within the body that operates much like a kingdom's defense against a tireless siege. This journey unveils the battle waged in the microscopic depths of the body, a constant conflict against an assault of microbial enemies that aim to invade and conquer. Each cell, each antibody, resembles a fortified castle or a skilled archer poised to protect the realm of human health.

The immune system is complex and wondrous; so here, one will break down its components and functions into something tangible and comprehensible. Think of it as learning how a castle withstands a siege: from the towering walls that repel invaders—the skin and mucous membranes—to the vigilant sentinels that patrol the ramparts—white blood cells and other immune components. This guide serves as a map to navigate the twists and turns of the body's innate and adaptive defense strategies without resorting to the perplexing jargon that often veils the clarity of science.

Within these pages, one will translate the intricacies of immunology into engaging, everyday terms. It's like understanding the strategies behind a legendary general's victory in battle—breaking them down into relatable tactics and maneuvers that not only inform but empower one with knowledge. The aim is to thread the needle finely between detail and digestibility, ensuring each sentence, each concept, interlocks in a coherent story of resilience and survival.

One will grasp the significance of every skirmish and learn how victories and defeats are met with ever-evolving strategies. Though the mechanisms are sophisticated and ever-changing, the narrative here is etched in transparency and simplicity, inviting one to step confidently through the curious and enigmatic world of immune responses.

Just as the high stone walls and imposing gates of a medieval fortress stand steadfast against invading forces, the body's protective barriers serve as the unsung heroes in our daily struggle against microbial marauders. Take the skin, for instance— a robust wall, resilient and weathered, which acts like a seasoned battlement, repelling unwanted advances with its sturdy ramparts. This outermost layer of defense is our personal armor, warding off pathogens with the casual might of an ancient stronghold.

Then there are the mucous membranes, akin to the fortress gates, vigilantly guarded and selectively permeable, allowing friendly travelers to pass while deterring the sneaky infiltrators. These living gates are equipped with their own traps and pitfalls—like sticky moats of mucus teeming with antibodies—that ensnare and neutralize would-be attackers before they can breach the inner sanctum.

Breathing life into these analogies is not mere fancy; it underscores the remarkable efficiency and nuance of the body's defenses. It's as if we're walking through life encased in our very own citadel, splendidly unaware of the daily sieges thwarted by our biological bulwarks. In understanding these first layers of defense in human terms—as walls and gates, guards and spears—one gains a newfound appreciation for the silent, steadfast watch that keeps us healthy and hale.

Here is the breakdown of the body's first line of immune defense, illustrated through the lens of a medieval battlement, where each component plays a crucial role in safeguarding the kingdom:

- **The Skin: The Fortress Wall**
  - **Epidermis**: The outermost barrier, like the sun-baked bricks of a castle wall, providing a physical impediment to invaders.
    - Keratinocytes: The sturdy stone blocks, filled with keratin, a tough protein that fortifies the barrier.
    - Langerhans cells: The vigilant sentries, ready to sound the alarm and initiate a defensive response upon spotting danger.
  - **Dermis**: Underneath lies the bastion's inner structure, home to the supportive connective tissue and the castle's water supply—blood vessels and lymphatics.
    - Fibroblasts: The dedicated engineers, maintaining the stronghold by producing collagen and elastin.
    - Mast cells and dendritic cells: Scouts and messengers, they detect breaches and relay messages to rally the immune army's reinforcements.

- **Subcutis**: The underlying cavernous storage, insulating the fortress and storing energy as fat.
    - Adipocytes: The fat cell storehouses, providing warmth and a reserve of energy to the castle's denizens.

- **Mucous Membranes: The Castle Gates**
    - The mucous membranes act like the gates of a fortress—discerning, allowing entry to allies and resisting foes.
    - Goblet cells: The oil for the gate's mechanisms, these cells produce the mucus that keeps the entry slippery and difficult to navigate for invaders.
    - Ciliated epithelial cells: They work like gatekeepers, wielding tiny brooms that sweep away trapped intruders in the mucus.
    - Immune cells within the membranes:
        - Macrophages: The quick-response knights, devouring any microbes that dare to attempt an infiltration.
        - Natural killer (NK) cells: The archers perched atop the walls, picking off targets that exhibit suspicious, pathogenic behavior.

- **The Mucus: The Moat**
    - The mucus is the moat around the fortress, teeming with defensive agents.
        - Antimicrobial peptides: The piranhas of the moat, these molecules actively chew through the walls of invaders.
        - IgA antibodies: The patrolling crocodiles, selectively binding to foreign entities and preventing their adherence to the walls.

In exploring these defenses with language painted from the palette of the past, one gains insight into how the body's external barriers function in concert to protect the realm within from the onslaught of disease-causing microbes. Each layer, each cell type, each molecule is meticulously crafted, much like the defenses of a medieval fortress, showcasing not only the prowess of our bodily defenses but also the ingenuity inherent in our very nature.

Pathogens, the microorganisms which can cause disease, are as varied in type and tactic as the travelers one might encounter on a bustling city street. To understand their roles, think of them as unwanted visitors, each with a different method of disrupting peace in the city. Bacteria are the pickpockets, some harmless if ignored but others aggressively snatching health away, multiplying at a breakneck pace.

Viruses work like con artists, slick and deceptive, hijacking the city's resources—our cells—to replicate themselves. They're dependent on a host to survive, as they can't reproduce on their own. Fungi are akin to vandals, sometimes content with simply defacing outer structures like skin, or sometimes infiltrating deeper, creating more havoc.

Parasites, on the other hand, are the swindlers of the bunch, taking up residence in a host for long periods and often going unnoticed while they sap nutrients and energy. Each pathogen's method of infection is unique, just as a criminal's method of committing a crime varies. Bacteria may burst through the city's walls, while viruses might use stealth to slip past the gates; fungi find weak spots to seep through, and parasites are smugglers, hiding away to evade detection.

Understanding these pathogens depends on grasping the mechanics of how they interact with our body's systems, much like understanding a city's vulnerabilities requires knowledge of its infrastructure. By learning how bacteria resist antibiotics, or how viruses mutate to dodge vaccines, this knowledge doesn't just chart the landscape of infection—it illuminates the paths we can take to bolster our defenses and maintain the city's health. These insights equip a person, whether a beginner learning about pathogens for the first time or a veteran healthcare professional, with the power to uphold well-being in the face of these microbial threats.

Let's take a deeper look at the covert operations our microscopic invaders undertake to thrive and survive within us. Picture bacteria as tiny burglars, rapidly cloning themselves in dark alleys—these are binary fission, splitting into two with an efficiency that rivals factory production lines. Yet, when the law enforcement of antibiotics arrives, some craftier types whip out biochemical lock-picks, like enzymes that dismantle the antibiotics, or switch up their entry codes with mutations, becoming virtually unrecognizable—thereby sidestepping capture.

Turning to viruses, these are hackers of the cellular world. A virus sneaks into a cell like a phishing scam slips through your email filters, tricking the cell into believing it's a friendly message. Once inside, it hijacks the cell's manufacturing center, prompting it to crank out viral components instead of its usual cellular products, effectively turning the cell into a virus-producing factory before sending the new viral troops out to conquer further territory.

Fungi operate more like elusive street artists, finding their canvas on or in the body. They secrete enzymes to break down tissue—like spray paint dissolving away the wall's facade—and establish colonies, sometimes visible as unsightly blotches and sometimes hidden below the surface, constantly seeking moisture and warmth to let their art proliferate.

Then, there are the parasites, the espionage agents of the pathogen world. These agents embed themselves for the long haul like sleeper cells within host tissue, siphoning off resources, and often going unnoticed. They might cloak themselves in the host's proteins, like an undercover spy in a disguise, to avoid detection by immune surveillance for months or even years.

These organisms each boast a repertoire of tactics to interact with, outwit, and sometimes overpower the body's intricate defense mechanisms. By relating the immune system to a well-guarded city and pathogens to skilled infiltrators, one can grasp the delicate balance of power that defines the struggle for health. It's a narrative of continual adaptation, a testament to the importance of modern medicine, and a reminder of the stark ingenuity written into the very fabric of life. Understanding this interplay readies one not just for the battle against disease but enriches one's appreciation for the body's capacity to protect and heal itself.

Imagine a quiet, peaceful night at an ancient fortress when suddenly, an intruder is spotted scaling the walls. Just as the night watchmen are quick to sound the alarm, calling the guards to action, the innate immune system launches its immediate response to any microbial breach. Picture the patrolling immune cells as these guards, always alert, ever-ready to confront invaders with a flurry of activity the moment they're detected. Phagocytes, akin to the fortress's soldiers, mobilize swiftly, engaging and neutralizing the offenders with precision and efficiency. Inflammatory responses flare up like signal fires, rallying more defenders to the site of intrusion and marking the spot with redness and swelling, a beacon of the ongoing skirmish. This ancient, deeply-engrained reaction is the body's first line, much like the archers ready at their posts, firing volleys of arrows—here, antimicrobial peptides—to halt the invaders in their tracks. Understanding the innate immune response through this lens not only captures the immediacy and coordination of this biological guard but also spotlights its critical role as the stalwart protector of our health, always on duty to ensure the integrity of our bodily kingdom.

Here is the breakdown of the innate immune system's tactical defenses against the unrelenting siege of infection:

- **Phagocytes: The Infantry**

- **Macrophages**: Think of these as the fort's heavy infantry, patrolling territories and engulfing invaders upon sight, digesting them with lethal enzymes.
  - Engulfment: The act of surrounding and consuming pathogens, like a garrison capturing spies.
  - Cytokine production: Sending out chemical distress signals, rallying other immune cells to the site of infection.
- **Neutrophils**: The swift scouts of the immune system, quick to arrive at the battlefront, releasing a barrage of substances that kill pathogens, akin to archers with a volley of arrows.
  - Degranulation: Releasing antimicrobial enzymes and substances to create a toxic environment for invaders.
  - Netosis: The unique ability to cast out nets made of DNA to trap and neutralize pathogens.

- **Antimicrobial Peptides and Proteins: The Cavalry and Archers**

- **Defensins and Cathelicidins**: Riders who charge into the fray, disrupting the very structure of the microbes, like cavalry smashing through enemy lines.
  - Membrane disruption: Puncturing the pathogen's membrane, causing the contents to spill out, as siege weapons might breach castle walls.
- **Complement System**: A group of proteins acting as a coordinated archer unit, marking targets for destruction or forming a pore complex to break them down.
  - Opsonization: Coating pathogens to make them more apparent to phagocytes, like shining a spotlight on an infiltrator.
  - Membrane Attack Complex (MAC): Forming a lethal pore in the pathogen's membrane, effectively a precision strike that leaves the enemy defenseless.

- **Inflammatory Mediators: The Horn Blasts and Signal Fires**

- **Histamines**: Flares that increase blood flow to the troubled region, allowing more immune cells to reach the area rapidly, akin to signaling for reinforcements.
- **Cytokines**: Horn calls that modulate the behavior of immune cells, mediating their growth, movement, and activity.

- Chemokines: Specific cytokines that act as a guiding beacon, directing cell movements, guiding defenders to key battle locations.
- **Prostaglandins**: Mediators that intensify the inflammation response producing pain and swelling, much like a siren's call alerts the local town of impending danger.

Each brigade of this immune army plays a critical role in a well-orchestrated campaign to defend the body's realm. It's not simply a battle but a series of calculated moves, feints, and counterattacks in a war waged beneath the skin—each reaction, each cell type, each molecule has its counterpart in the grand theaters of human and historical conflict. By framing the immune system's response this way, the profound sophistication behind the simple act of staying healthy becomes as clear and appreciable as enjoying a good story over a cup of coffee with a friend.

Imagine the adaptive immune system as a guild of elite knights, each specially trained for a unique kind of quest against specific adversaries. Upon encountering a threat, these knights — B cells and T cells — embark on their mission with singular focus. B cells, like archers equipped with an arsenal of bespoke arrows (antibodies), unleash a targeted volley toward the enemy that is marked for destruction. Each arrow is custom-forged and uniquely designed to fit the intricate locks of their targets, neutralizing invaders with remarkable precision.

Meanwhile, T cells operate in two specialized squads. The first, the Helper T cells, act as battlefield strategists, amplifying the guild's defenses by directing the B cells and rallying their defensive efforts. The second, the cytotoxic T cells, function as the knights who bear close-quarters combat weapons that deliver lethal blows, destroying infected cells with the accuracy and finesse of master swordsmen.

This deployment is no random sortie but rather a calculated engagement; the immune system's memory cells are like scholars who remember the faces of past foes, ensuring that the knights are better prepared for any future encounters. To witness these knights in action is to observe a symphony of defense—a masterclass in strategy and execution that epitomizes the body's remarkable capability to adapt and overcome. Engaging with this scenario lets us appreciate the beauty and intricacy with which our bodies protect us, much like relishing an epic tale of chivalry and courage shared among friends.

In the adaptive immune response, the body acts like a covert intelligence network, gathering data on specific pathogens to mount a precise counterattack. B cells and Helper T cells are the front-line operatives in this network. B cells can recognize invaders directly, while Helper T cells require a liaison—a professional presenter of sorts. This liaison, often a professional antigen-presenting cell such as a dendritic cell, displays pieces of an invader, known as antigens, on its surface, essentially putting up "Wanted" posters that Helper T cells can read.

Once the Helper T cells are briefed, they send signals, much like flares, calling for backup. This is when clonal expansion takes place, calling on the B cells to produce copies of themselves, specifically designed to target the identified enemy. Think of these B cells as a specialized unit trained for that sole purpose, now equipped and ready to seek and destroy their target.

Antibodies, produced by the mobilized B cells, work like homing missiles. Their tailored structures allow them to lock on precisely to the antigens presented by the enemy, flagging them for removal or directly neutralizing them. Their precision in finding their targets in a sea of bodily cells is akin to a GPS-guided weapon system, ensuring that the right targets are taken out without collateral damage.

Cytotoxic T cells are the special forces in this operation. Once activated, they search out the cells infected by the pathogen. Upon finding their target, they engage in direct combat, delivering a lethal hit that causes the infected cell to self-destruct. This process, known as apoptosis, can be likened to placing a precisely set explosive charge that only destroys the compromised infrastructure.

Lastly, memory cells are like seasoned veterans or a wise council that keeps detailed records of each invader the body has ever encountered. They remain in the body long after the battle, ready to recognize and respond faster if the invader returns. This means subsequent responses are more rapid and effective, a phenomenon that underpins the concept of immunological memory and the basis of how vaccines work.

Understanding the adaptive immune system through this narrative reveals the sophistication behind how the body identifies, targets, and remembers pathogens. It's a testament to the natural world's complexity and the ingenious ways in which organisms protect themselves—an eloquent dance of biology that keeps us alive and healthy amidst a continuous barrage of microbial threats.

When a pathogen manages to slip past the body's sentinels and starts a successful infection, the immune system swings into action, much like emergency services responding to a crisis. Initially, the battle against the invader might cause symptoms like fever or inflammation, the body's equivalent to road flares and sirens—it's a sign that help is on the way, and restoration efforts are in full swing. During this time, white blood cells converge at the infection site, targeting and destroying the marauders in a meticulous clean-up operation.

As the tide of battle turns and the pathogens are cleared, the body shifts from defense to repair mode, undertaking a process similar to a construction crew repairing damage after a storm. Specialized cells, such as fibroblasts, start rebuilding tissues while growth factors and hormones work in tandem to promote healing, effectively acting as foremen overseeing the repair project. The immune system, meanwhile, takes stock of the incident, creating memory cells that act like city planners who make note of what happened to prepare better for any similar events in the future.

If the infection was significant, healing may take time; it's not a rushed patch-up job. The body carefully orchestrates recovery, mending its tissues layer by layer until integrity is restored. This intricate process, hidden beneath the surface, is essential to regaining full function and preventing future vulnerabilities.

Throughout recovery, it's crucial to support the body with adequate rest, nutrition, and, if needed, medical intervention to ensure the healing process proceeds without a hitch. Just as cities rebuild and fortify after setbacks, the body emerges stronger and more prepared from each encounter with disease, a testament to its remarkable resilience and the immune system's pivotal role in health and recovery.

When a pathogen breaches the body's initial defenses, it's akin to an alarm being tripped in a city's high-tech surveillance system. Neutrophils are often the first responders on the scene, like the city's police force, arriving quickly to engulf and neutralize the threat through a process called phagocytosis.

These foot soldiers also release chemicals that act like flares, signaling the magnitude of the invasion to other immune cells.

Macrophages are like the heavy artillery and clean-up crew combined. They not only ingest pathogens like neutrophils but also dispose of dead cells and debris, and they produce a wider array of signaling molecules. These signals, much like emergency broadcasts, can induce inflammation; they help wall off the affected area, creating a barrier to prevent further spread, similar to a city's lockdown.

Dendritic cells act as the intelligence officers. They gather pieces of the pathogens - like evidence collected at a crime scene - and present them to T cells in the lymph nodes. This is a critical step, as it activates the adaptive arm of the immune system, calling in specialized units for reinforcement.

During the inflammatory response, blood vessels widen to allow more immune cells to access the infected area, akin to opening roads for emergency vehicles. This rush of cells to the site is guided by chemical signals - cytokines and chemokines - drawing them to the source of infection just as a GPS system guides responders to the right location.

As the immediate threat is contained and neutralized, fibroblasts start the repair process. These cells lay down new connective tissue to replace what has been damaged, a bit like construction workers repairing roads and buildings following a disaster. They're supported by growth factors, akin to city planners and engineers who ensure the repairs are structurally sound and up to code.

Amidst these repairs, memory B cells and T cells form. They're like historians and archivists, retaining detailed records of the pathogens so if the same type attempts to invade again, the city - our body - can recognize and respond to them more swiftly and vigorously.

This entire process from detection to recovery showcases the body's advanced, multi-layered immune response, reflecting a resilience and capacity for self-healing that mirrors the emergency response and reconstruction efforts of a well-organized city after an unforeseen event. The analogy captures the essence of each immune element's part in this complex and finely tuned system, illustrating the beauty and sophistication of the body's way of protecting and maintaining itself.

Imagine vaccines as a series of elaborate defense drills for a city's armed forces. In peacetime, the city's leaders don't rest; instead, they prepare for potential threats by simulating invasions. Vaccines work in much the same way; they train the body's immune system, presenting it with a harmless piece of the enemy — just enough to recognize and remember, but not enough to cause harm. It's like a mock battle where soldiers practice their strategy so that when the real threat arrives, they're ready to respond with swift precision.

In contrast, when an infection strikes and the city is under siege, drugs come in like elite reinforcements parachuting in at the eleventh hour. Antibiotics and antivirals act as specialized units to support and boost the beleaguered troops — the immune cells — helping them gain the upper hand against live invaders.

These medical interventions, the rehearsals and last-moment rescues, are not just about surviving the immediate threat; they're about fortifying the city's defenses, ensuring its future safety. With vaccines eliciting long-term preparedness and drugs delivering immediate aid, the city — our body — becomes all the more impregnable. emlrtnews

Here is the breakdown on how vaccines and drugs serve as vital reinforcements for the body's immune brigade:

- **Vaccines: Training Simulations for the Immune System**
    - **Live-attenuated vaccines**: Like battle reenactments with blunted swords, these use weakened forms of the virus to practice defense without the risk of a full-blown attack.
    - Example: Measles vaccine.
    - B and T cell response: These vaccines stimulate a robust immune reaction, leading to long-lasting immunity.
    - **Inactivated vaccines**: Utilizing inactive, but intact, pathogens to train the immune system, akin to drills with a dummy opponent.
    - Example: Polio vaccine (given as an injection).
    - B and T cell response: They prompt an immune response focused on the pathogen's recognizable features.
    - **Subunit vaccines**: Providing only essential components of the

pathogen, like showing the immune system the enemy's flag without confronting the enemy itself.
  - Example: HPV vaccine.
  - B and T cell response: These target very specific parts of the pathogen to prevent infection.
  - **mRNA vaccines**: They act as a high-fidelity training program, instructing cells to produce a protein unique to the pathogen, effectively staging a realistic yet controlled mock invasion.
  - Example: COVID-19 vaccines from Pfizer-BioNTech and Moderna.
  - B and T cell response: These vaccines lead to the production of protein pieces that stimulate a targeted immune response.

- **Drugs: Critical Reinforcements during Infections**
  - **Antibiotics**: These are the elite demolition experts, specifically designed to breach the cell walls of bacteria or block their vital functions, like shutting down the enemy's supply lines.
  - Mechanism: Targeting bacterial ribosomes or cell wall synthesis.
  - Immune system augmentation: They reduce the bacterial count, giving the immune system an upper hand.
  - **Antivirals**: Covert operatives that interfere with a virus's ability to replicate inside host cells, akin to jamming the enemy's communication lines.
  - Mechanism: Hindering viral entry, replication, or assembly.
  - Immune system augmentation: They directly limit viral spread, allowing the immune system to clear the infection more efficiently.

Understanding these medicines is akin to grasping the various strategies of an army preparing for and engaging in battle. Vaccines act as drills to prime the immune system for potential threats, building a formidable line of defense before an enemy strikes, while antibiotics and antivirals provide immediate tactical support to combat active incursions. This ensures a strong, strategic immune response, keeping the 'city'—our body—safe and sound.

Consider the body as a fortress in a perpetual state of readiness, where one's lifestyle is the daily drill that keeps the castle garrison, the immune system, in top form. Just as a well-maintained fortress can withstand sieges, a body fueled by nutritious food, engaged in regular physical activity, and rested by ample sleep forms a formidable barrier against potential illnesses. Nutrition, akin to the supply chain for a castle, delivers vital provisions—vitamins, minerals, and energy—that feed and fortify the troops. Exercise, much like routine patrols, keeps the soldiers alert and walls reinforced, enhancing circulatory health and bolstering defenses.

Sleep, arguably one of the most underrated strategies in maintaining a stalwart defense, acts like the repair and rejuvenation of the castle walls at night. When sleep is short-changed, the ramparts might crumble more quickly under attack. Similarly, when one is well-rested, the cells can efficiently repair and the body can better ward off invaders.

Stress management also plays a key role; one can liken it to the system in place to keep a lookout for any signs of unrest among the garrison. Chronic stress can be analogous to a spy that weakens the fort from the inside, compromising its defenses.

One's lifestyle choices are not just about individual wellbeing; they are also part of the collective resilience against public health threats. Just as a fully operational fortress protects an entire region, a healthy individual contributes to the strength and immunity of the community. By maintaining a healthy lifestyle, the body ensures that its defenses are ever-vigilant and ready for whatever challenges may emerge.

Let's take a deeper look at the intricate ways in which lifestyle choices weave into the tapestry of our immune system's strength and efficiency. Think of vitamin C as the blacksmith within the fortress, tirelessly forging the weapons—antibodies—needed for defense. Vitamin D acts much like the training instructor for new recruits, bolstering the killing capacity of immune cells like macrophages, while zinc works behind the scenes to maintain communication lines between cells, ensuring a coordinated defense.

As for exercise, picture aerobic workouts as routine patrols around the fortress perimeter, boosting surveillance—the circulation of immune cells—while anaerobic exercise builds the fortress walls—strengthening the cellular components of immunity. Combined, they keep the soldiers fit for battle and the fortress itself structurally sound.

Sleep, the grand strategist of the fortress, orchestrates nightly repair strategies that ensure the defense is at the ready come dawn. This crucial act promotes the release of cytokines, which signal and direct the immunological troops to areas of need. On the flip side, a command center operating without rest—chronic sleep deprivation—scrambles these signals, leading to gaps in the fortress's defenses.

And stress, akin to a traitor within the castle walls, invites havoc by continuously sounding false alarms—persistently releasing cortisol—which eventually wears down the alertness of the troops and can diminish the fortress's integrity.

By understanding the minutiae of these lifestyle factors, one appreciates that they are not merely a checklist for good health but are integral components of an elaborate defense strategy, ensuring that the fortress stands resilient against all adversities that may come its way.

Amid an unending stream of microbial threats, the immune system stands out as an indefatigable protector, tirelessly patrolling and defending the body's landscapes. It is a testament to biological ingenuity, a complex network of cells, organs, and proteins that together, construct a fortress against the microscopic marauders that relentlessly test our defenses. Each battle won—in silence, without our awareness—is a marvel, showcasing the immune system's remarkable ability to adapt and learn from each new challenger it meets.

The upkeep of this internal bastion is crucial—one's own habits and lifestyle choices act as the maintenance and training regimen that keeps these microscopic sentinels sharp and prepared. A balanced diet, rich in essential nutrients, assures that the immune cells are well-fueled for their vigilant patrol. Consistent physical activity ensures that these cells circulate efficiently through the body, like scouts surveilling the terrain. Adequate sleep acts as the critical downtime that allows for the repair and recalibration of immune defenses, making certain they are ready for the next day's challenges. And stress management is the tactic of keeping a clear head in the face of potential false alarms so that the system does not become overtaxed and thus, more susceptible to an actual invasion.

Understanding the role of these simple, daily actions in bolstering the immune system's readiness is to grasp the fine art of a seamless interplay between lifestyle and cellular function. Maintaining the immune system in a state of readiness is not just an act of personal well-being, but like the brushstrokes on a grand canvas, it contributes to the collective picture of human health. It is an ongoing effort, subtle but profound, an investment in one's body to ensure resilience in the face of microscopic challenge.

INVISIBLE ENGINEERS THE ENVIRONMENTAL IMPACT OF MICROBES

Picture, if you will, the bustling metropolis of a city – each building meticulously designed, each service finely orchestrated, and each structural element critically placed. Now, imagine if all this complex organization and design work were the doing of masterminds unseen to the naked eye. This is precisely what unfolds in nature's grand scheme, where microbes are the elusive architects, crafting the life-sustaining functions upon which our planet depends.

Microbes engineer the city of life by cycling nutrients, purifying water, enriching soil, and creating oxygen – acts as essential to Earth's well-being as the foundations are to a skyscraper. Without the continuous work of these microscopic marvels, the globe's ecosystems would grind to a halt, like a city deprived of electricity or water. Steps in understanding this invisible workforce start by unraveling how these organisms are indispensable to the environment – renewing it, maintaining its balance, and enabling the rich tapestry of life that thrives upon it.

This opening dialogue aims to unravel the complexities of microbial activity, breaking it down into its most fundamental components, much like dividing a city into neighborhoods, roads, and utilities. An understanding of these tiny environmental artisans helps appreciate their grand constructions, revealing, layer by layer, their pivotal role in Earth's biosphere. While they may seem to work in shadow, their impact shines brightly on the surface, for it is their ceaseless diligence that nourishes the planet, sustaining life in all its diversity.

Just as the Roman aqueducts stand as a testament to masterful engineering, channeling life-giving water to cities, so too do microbes orchestrate the flow of life-sustaining nutrients through the environment. Consider them the skilled craftsmen of the microscopic world, diligently shaping and constructing the complex bio-networks beneath our feet. These micro-engineers labor tirelessly, unseen but as pivotal as the ancient workers who pieced together the Great Wall, stone by stone, to protect and sustain their civilization.

Each microbe plays its part, like a stonemason or architect, contributing to a greater design. Some harness the power of the sun, building food from light and air, while others break down waste, turning what was once discarded into treasured resources. Together, they create an infrastructure that supports life on a global scale, as magnificent and essential as the great marvels of human construction. Only, their monuments aren't meant for the sky but for sustaining the vibrant tableau of life that exists on this planet. Through this analogy, we can begin to see how these unseen micro-constructors build and maintain the awe-inspiring edifice that is our natural world.

Here is the breakdown of the diverse roles microbes play as ecological engineers, woven into an intricate tapestry of life-supporting processes akin to the vast network of utilities and services in a functioning city:

- **Photosynthetic Microbes:**
    - Like solar panels harnessing sunlight, these microbes, such as cyanobacteria and algae, capture energy from the sun and convert it into a form that sustains food chains.
    - Sublist:
        - Oxygen Production: They release oxygen as a byproduct, similar to how a power plant provides electricity to light up homes.

- **Decomposers:**
    - Comparable to recycling centers, decomposer microbes break down dead organic material, returning nutrients to the environment.
    - Sublist:
        - Soil Fertility: They help maintain soil fertility, like a plant converting waste into compost for gardening.
        - Waste Reduction: Their ability to decompose reduces environmental waste, akin to technological innovations in waste management.

- **Nitrogen-Fixing Bacteria:**
    - These bacteria act like natural fertilizer factories by converting atmospheric nitrogen into a form that plants can absorb and use.
    - Sublist:
        - Increase Crop Yield: Much like fertilizing a field, they enhance the soil, thereby boosting plant growth and food production.

- **Symbiotic Relationships:**
  - Microbes engage in symbiotic relationships with plants, like trade agreements between businesses, each providing vital services to the other.
  - Sublist:
    - Root Health: They support the health of plant roots, reminiscent of constructing a robust foundation for a building.

- **Water Purification:**
  - Certain microbes have a role akin to water treatment plants, filtering and purifying water, ensuring that rivers and lakes sustain aquatic life.
  - Sublist:
    - Detoxification: They detoxify pollutants, similar to sophisticated water filtering technologies that remove impurities.

- **Climate Regulation:**
  - Microbial activities influence climate by regulating greenhouse gases, drawing parallels to climate control systems that manage temperature and air quality.
  - Sublist:
    - Carbon Sequestration: Some microbes sequester carbon, functioning like technologies developed to capture and store carbon emissions.

- **Biodiversity Preservation:**
  - By performing their varied functions, microbes contribute to the maintenance of Earth's biodiversity, like architects designing a city with green spaces for wildlife.
  - Sublist:
    - Habitat Support: Microbes create and sustain various niches and habitats, crucial for the survival of innumerable species.

This clear and accessible explanation of microbes' roles showcases their unseen yet monumental impact. As a knowledgeable companion, this narrative invites an appreciation for the silent yet profound work of these invisible engineers, casting light on the connective threads between their microbial efforts and the vast web of life they support.

Nitrogen fixation is like a translation process where certain microbes – think of them as specialized alchemists – convert nitrogen, an abundant but inert gas in the air, into a form that plants can use to grow. Plants can't use nitrogen directly from the air; it's like trying to listen to a radio without turning it on. These microbial alchemists grab that nitrogen, turn it on, so to speak, and change it into a nutrient, almost like turning air into food.

Meanwhile, the carbon cycle involves different types of microbes that act as both producers and decomposers. Producers, such as plants and algae, take carbon dioxide from the air and, using sunlight, turn it into food via photosynthesis – consider them the chefs who prepare meals from raw ingredients. Decomposers are like cleanup crews that take dead plants and animals and break them down, releasing carbon back into the atmosphere as carbon dioxide, keeping the cycle going.

Other biogeochemical cycles work similarly, with microbes fulfilling roles of transforming substances from one state to another, enabling elements like phosphorus and sulfur to be used by plants and animals. It's a relay race of chemistry and biology where microbes pass the baton of life's essential elements, ensuring that nothing goes to waste and everything is in continuous motion, supporting all forms of life on Earth.

Approaching these topics without technical jargon isn't just about simplicity; it's about providing a clear picture of the invisible yet indispensable work of microbes. By understanding their roles, one gains not just knowledge but a deeper respect for how these tiny organisms shape the world.

Let's take a deeper look at the world beneath our feet, where microscopic workhorses play a role as vital as any gardener tending to their plots. Nitrogen-fixing bacteria like Rhizobium and Azotobacter are the unsung heroes in the plant world. They cozy up to the roots of legumes in a friendly association where both parties benefit—like having a roommate who pays rent in a currency that plants love: ammonia. These bacteria take nitrogen gas, which makes up most of the air we breathe but is as unusable to plants as a locked treasure chest, and convert it into a form as accessible as a key, unlocking plant growth and nourishment.

Moving on to the carbon cycle, imagine autotrophs—plants and algae—as the chefs in nature's kitchen, whipping up glucose in the grand cook-off known as photosynthesis. They take in carbon dioxide, mix it with water, and, under the sun's watchful gaze, create their meal and oxygen—a byproduct that's like the delightful aroma wafting from the kitchen, essential for our very breath. Heterotrophs, including microbes, animals, and even us humans, are the diners feasting on these glucose-rich meals, and through respiration, we complete the cycle, exhaling carbon dioxide back into the environment.

But the plot thickens as these cycles intertwine with the phosphorus and sulfur cycles. Phosphorus is like the currency of energy in biological systems, vital for DNA and cellular energy, while sulfur is a key player in proteins, akin to the rebar in concrete, providing strength and structure. Microbes facilitate the movement of these elements through the environment, transforming them into various compounds—some soluble, others not—much like alchemists transmuting metals, ensuring that these elements are in the right form and the right place to be used once again by plants or other organisms.

This intricate dance of transformation and energy flow underscores the significance of each cycle and the microbial maestros orchestrating it all. With this detailed breakdown, like understanding the inner workings of a clock, we can better appreciate the precision and balance necessary to sustain life's symphony. So the next time you marvel at lush greenery or take a breath of fresh air, remember the tiny giants laboring in obscurity to hold up our world.

Imagine walking into a vast, echoing hall. This hall, resembling our water systems, can sometimes be overtaken by an army of algae when conditions are just right—much like an unattended banquet hall might attract a flock of opportunistic pigeons. Algae blooms, the result of this biological gathering, can disrupt the delicate ecosystem of the water, depleting oxygen and blocking sunlight, akin to how a thick crowd can make a room stuffy and dim. It's an example where too much of a good thing turns troublesome, like overwatering a plant.

Now, switch the scene to an unfortunate spillage—a slick of oil darkening the surface of the ocean. Here come microbes to the rescue, serving as nature's cleanup crew. These microscopic custodians take on the oil, breaking it down bit by bit as diligently as janitors who tackle a disruptive stain on a marble floor. They work quietly, without fanfare, but their efforts can transform a devastated waterway back into a thriving habitat, much like a carefully restored work of art.

These case studies aren't just tales of calamity and recovery; they're real-life demonstrations of the role microbes play as both the inadvertent disruptors and the unsung heroes in our environment. Through understanding how these tiny organisms can create and solve crises, one can grasp the scale of their influence and the vital importance of keeping their populations in balance, ensuring the enduring health of our planet's water systems.

Let's take a deeper look at the aquatic troublemakers known as algae blooms, akin to a rowdy crowd that crashes a calm neighborhood block party, transforming it from peaceful to chaotic. The species often responsible include microorganisms like **Microcystis**, **Anabaena**, and **Pseudo-nitzschia**, notorious for their ability to explode in numbers when the mix of warm water, abundant sunlight, and rich nutrients is just right—much like the perfect party conditions attract more guests. These conditions often stem from excessive nutrient runoff from agriculture, mirroring how too many invitations can lead to an overcrowded house.

The aftermath of such blooms is dire, much like the mess left behind after a big bash. They create hypoxia, a severe oxygen shortage in water, akin to a room so packed with people that it becomes stifling and hard to breathe. This depletion of oxygen wreaks havoc on marine habitats, causing fish to suffocate and die, not unlike partygoers desperate for fresh air.

On to the scene of an oil spill, nature dispatches its microscopic cleanup brigade, including the specialized bacteria **Alcanivorax borkumensis**. These microbes are the unsung heroes of the seas, their appetite for oil rivaling a locust's hunger for crops. They secrete enzymes, specialized molecules that act like a set of tools, to break down the complex hydrocarbons in oil into simpler, harmless compounds, effectively cleaning up the spill bit by bit, just as a team might tackle a daunting cleanup after a festival.

Their metabolic pathways, the routes taken by nutrients inside these bacterial cells, are like intricate highways that process and convert food into energy—highways that, when functioning at full capacity, can help restore polluted waterways to their pristine state. Think of these microbes as both waste processors and healers of the marine world, capable of turning a marine disaster site back into a vibrant aquatic neighborhood, where life continues and thrives.

Understanding these tiny laborers gives new meaning to the phrase 'small but mighty', as their silent work beneath the waves plays a crucial role in maintaining the health and balance of our oceans. Through this exploration, one can better appreciate the critical functions these microorganisms perform to keep aquatic ecosystems—our planet's blue heart—beating strong.

Human actions, such as the use of antibiotics and certain agricultural practices, can have profound effects on the microscopic communities that underpin our environment. Take antibiotics, for instance. When they find their way into natural settings, either through medical overuse or agricultural runoff, it's akin to dropping a bomb in a city, indiscriminately hitting both harmful and helpful inhabitants. This assault can disrupt the delicate balance of microbial life, often leading to the rise of antibiotic-resistant species – the toughest of the bunch, much like weeds that thrive after a forest fire.

Agriculture, particularly when it comes to the use of fertilizers and pesticides, is a double-edged sword. On one side, these practices can bolster crop production, akin to giving a runner energy-boosting supplements. However, excess chemicals can seep into nearby ecosystems, acting like uninvited guests that crash the natural party, throwing off the balance and harmony among resident organisms.

These practices can lead to changes in environmental health, such as eutrophication in water bodies – a phenomenon similar to overfeeding fish in a pond, resulting in a glut of algae that sucks life-giving oxygen out of the water. In the soil, they can alter microbial compositions, akin to changing the line-up of a sports team, affecting its performance and, subsequently, the health and diversity of plant life that depends on these underlying microbial functions.

Each swig of antibiotic medication, each sweep of pesticide, sends ripples through the complex web of life, reminding us that every action has a reaction, often far more reaching than one might expect. By understanding these connections and consequences, individuals can make informed decisions, appreciate the hidden yet critical world of microbes, and become stewards of a more resilient and diverse environment.

When antibiotics leach into ecosystems, say from a farm runoff or improper disposal, they start a domino effect that disrupts the microbial community. This process begins with the antibiotics entering waterways, much like unwanted chemicals slipping into a city's water supply. They encounter a diverse world of bacteria, where some are susceptible to the antibiotic and die off, while others, sporting natural resistance, survive. These survivors, now with less competition, flourish and multiply, passing on their resistant genes as effortlessly as a viral trend.

Meanwhile, agricultural chemicals like fertilizers wash into rivers and lakes, setting the stage for nutrient overload. This scenario is similar to over-fertilizing a garden, causing explosive weed growth that chokes out other plants. In water bodies, the excess nutrients, notably phosphates and nitrates, become a feast for algae, leading to massive blooms. As these algae die, they decompose, a process that consumes oxygen and releases yet more nutrients, propelling the cycle. This can create 'dead zones' akin to deserted urban districts, where the absence of life is stark and suffocating.

Microbes, ranging from nitrogen-fixing bacteria to phosphorus-unlocking species, find their roles and numbers skewed in these altered environments. Relationships that once formed webs of mutual benefit, like well-orchestrated city services, now fray and unravel, threatening the biodiversity and resilience of the ecosystem.

To counteract these impacts, strategies akin to urban renewal projects can be employed. Practices like using fewer antibiotics, applying them more judiciously, and managing agricultural runoff via buffer zones mirror sustainable urban development that can revitalize a city. Phasing out harmful chemicals, encouraging organic farming, and implementing precision agriculture can restore balance to the environmental 'cityscape,' ensuring that every microbe and nutrient finds its rightful place in nature's grand design.

In the vast and intricate tapestry of life, every thread counts, including the fine, often invisible threads woven by microbial life. Reflecting on the stride that conservation efforts have made, reducing pollutants like industrial waste and agricultural run-off isn't just beneficial; it's essential. These steps are much like cleaning a river that's become choked with refuse – where the removal of each piece of litter can slowly but surely clear the waters and allow life to return in thriving abundance. Protecting natural habitats, ensuring that forests, wetlands, and coral reefs remain intact, safeguards these tiny architects of Earth's biogeochemical cycles and stewards the vast genetic library that microbes have curated over billions of years.

Cutting down on pollutants helps prevent episodes of environmental damage, such as the dead zones in water bodies – where, akin to barren wastelands, nothing can live. This preserves not only the majestic creatures that captivate our imagination but also the unseen microbial communities that support them. Reducing habitat destruction has an equivalent effect in conserving microbe diversity, much as retaining old manuscripts keeps historical knowledge alive for future generations.

In making strides to safeguard the microbial diversity, humans can ensure the wheels of these natural cycles continue to turn smoothly, like cogs in a grand clockwork. Preserving this micro-diversity is vital, for it is the bedrock upon which the health of the planet – and by extension, the future of all life, including ours – rests. It requires diligence, the ability to look beneath the surface, and the wisdom to recognize that when it comes to life's diversity, even the smallest beings have an enormous role to play.

Think of each eco-friendly product you choose like a vote for the kind of world you want to live in. Just as picking up litter during a morning jog leaves the trail cleaner for those who follow, opting for products with reduced packaging or made from sustainable materials creates less waste and pollution. It's a simple switch from plastic to a reusable coffee cup, but imagine if that tiny act were multiplied across every morning brew in the city—suddenly, you've got a movement that lightens the load in landfills.

Supporting responsible waste management can be compared to forming a neighborhood watch. It's being part of a community that's mindful of where its trash ends up, much like neighbors keeping an eye on each other's homes. When you compost, recycle, or properly dispose of hazardous materials, it's like ensuring the streets stay safe; you're helping to keep the environment clean and reducing the chance of pollution causing harm. Each of these choices may appear small, like choosing to water your garden with a watering can instead of a hose, but over time, and with enough gardeners, that water saved could fill a lake.

Upon reflection, the magnitude of change that can be achieved through collective individual actions is akin to a mosaic — each eco-conscious decision is a single tile, and together, they form a larger picture of a sustainable future. It is this very approachability and sense of community in our everyday choices that hold the power to educate, inspire, and effect lasting environmental change.

Let's take a deeper look at the ripple effects a single eco-friendly decision can set into motion. Choosing a reusable coffee cup over disposable ones might seem trivial, but the average person's yearly use can pile up to about 365 cups—that's a stack taller than a two-story house! Now, if a whole neighborhood makes the switch, that's fewer trees cut down for paper, less plastic coating in landfills, and a significant dent in the stream of pollutants from manufacturing.

Diving into composting, think of it as giving back to the earth, quite literally. Kitchen scraps and yard waste can transform into rich, nutritious soil, much like a chef creates a gourmet meal from raw ingredients. This back-to-basics tactic not only enriches the soil, helping plants to flourish as if they were guests at a five-star all-you-can-eat buffet, but it also reduces methane emissions from landfills—those silent climate change contributors.

In the world of recycling, every aluminum can or glass bottle spared from the trash is a mini victory. Recycling aluminum saves about 95% of the energy it would take to make new metal from ore, the equivalent of turning off a major city's power for a whole day with just one can's energy savings. Glass recycling is just as impact-rich, saving raw materials and slashing pollution, all while the items retain their quality, unlike a game of telephone where the message distorts with each pass.

Turning to waste management, properly disposing of hazardous materials means keeping toxins out of our waterways and away from wildlife, like removing traps from a forest path. It's about ensuring that chemicals don't linger, ready to seep into water sources, like uninvited guests lingering long after a party ends, with the potential to cause chaos.

When people come together, making these mindful decisions part of a collective routine, their compounded effect is monumental. Just as a beehive's strength is the sum of its tiny workers, collective eco-friendly actions forge a path to sustaining our planet's resources and curbing the juggernaut that is climate change. By considering the future impact of present choices, each person becomes not just a consumer but a caretaker of the environment, stewarding us toward a greener tomorrow.

In closing, consider the microbe: a world in a drop of water, a universe in a speck of soil. These tiny but mighty entities are the unsung champions of Earth's resilience, operating like intricate cogs in the vast machinery of global processes. Their unseen actions—breaking down waste, recycling nutrients, purifying water—serve as the fundamental operations for life as we know it. Microbes are the virtuoso conductors of the symphony of life, ensuring every note resonates in harmonious balance, from the depths of the oceans to the canopies of the rainforests.

Without these microscopic marvels, the planet's health would falter, much like a city plunges into chaos without its key services. The breath of life itself relies on their continual labor: they are the original life-support system, functioning tirelessly, generation after generation. Their role extends beyond merely sustaining the present; it encompasses the tireless shaping of the future, ensuring the legacy of biodiversity for ages to come.

Acknowledging their pivotal role reframes one's perspective on the natural world, recognizing microbes not as mere inhabitants of the biosphere, but as its very architects, trusted with the blueprint of life's complexity. So, as this chapter concludes, let it leave behind an enduring thought: to understand and protect these foundational elements of our world is to embrace the true essence of our interconnected existence, paving the way for a future as vibrant and lively as the microbial dance upon which it rests.

MICROBIAL HARMONY SYMBIOSIS IN THE MICROBIOME

Step into the hidden realm of the microbiome, a place where trillions of microscopic organisms inhabit your body, creating a complex and vibrant community much like a teeming rainforest within. Here, life flourishes in a delicate dance of give-and-take, where each microscopic inhabitant plays a critical part in maintaining the health and stability of its vast ecological network, the human host. This chapter is an open invitation to explore these microscopic synergies — not through a microscope, but through the lens of everyday experience.

As a guide through this unnoticed world, the goal is to unravel the intricate relationships that form the tapestry of life within, teasing out the threads so that anyone can grasp their significance. You'll discover how these tiny life forms have a giant impact, aiding in everything from digesting the morning toast to safeguarding against diseases. It's a journey that showcases not just the incredible adaptability and resilience of the human body, but also the wonders of the organisms that dwell within, interwoven with our existence more intimately than one might have ever imagined.

By the end of this chapter, the hope is to leave you with more than knowledge; it's to inspire an appreciation for these diminutive allies and a recognition of the vital role they play in our lives. So, take a breath, and prepare to be astonished by the microscopic marvels that thrum beneath the surface of your skin and within the depths of your gut—for in the world of the microbiome, small is mighty, and the quietest whispers can resonate with the profoundest effects.

Imagine walking into a bustling coffee shop, where the air is rich with the scent of brewed coffee and the sound of gentle jazz. Here, you find symbiosis in every corner: the barista crafting your perfect cup hinges on the farmers who harvested the beans, much like flowers rely on bees for pollination. This is mutualism, where each party thrives from their interaction, like dancers flawlessly in step with their partner's rhythm.

In the corner, a charging station provides power to patrons' phones without needing anything in return. This is much like commensalism in the microbiome, where one organism benefits while the other remains unaffected, similar to birds nesting in a tree that does not mind the extra weight.

Yet, just outside, a parking meter runs diligently, demanding payment for a spot on the street. This is akin to parasitism, where one benefits at the expense of the other, reminiscent of a ticket slapped on a windshield — a one-sided relationship that leaves one party feeling a bit shortchanged.

These everyday exchanges reflect the delicate balance of symbiosis within our own bodies, where microbes engage in a constant give-and-take. Each relationship, whether mutualistic, commensal, or parasitic, plays a part in the grand dance of life, contributing to a dynamic, interconnected system where every player has a role vital to the health and wellbeing of the whole. It's a harmony of existence, choreographed to the subtle music of survival and coexistence.

Here is the breakdown on the intricate ballet of symbiotic relationships within the human microbiome:

- **Mutualistic Relationships:**
    - **Microbes Involved**: Beneficial bacteria like Bifidobacteria and Lactobacillus.
    - **Nutritional Exchange**: Bacteria ferment dietary fibers that the human digestive system cannot process, breaking them down into short-chain fatty acids (SCFAs).
    - **Physiological Outcomes**:
    - Host Benefits: These SCFAs nourish the intestinal cells, reduce inflammation, and may help protect against colon cancer.
    - Microbial Benefits: The bacteria thrive on the fibers, enabling their growth and continued residence within the gut.

- **Commensalistic Relationships:**
    - **Microbes Involved**: Skin flora such as Staphylococcus epidermidis.
    - **Nutritional Exchange**: Bacteria consume sweat, sebum, and skin cells, sourcing nutrients without affecting the host.

- **Physiological Outcomes**:
- Host Benefits: The skin remains unaffected while benefiting from the prevention of pathogenic colonization, akin to birds that eat pests off the backs of buffalo.

- **Parasitic Relationships:**
- **Microbes Involved**: Harmful bacteria, viruses, or fungi such as Streptococcus pneumoniae or influenza viruses.
- **Resource Depletion and Toxins**: These pathogens may consume host resources, produce waste, or release toxins.
- **Physiological Outcomes**:
- Host Consequences: This can lead to symptoms or diseases, as the pathogen thrives at the host's expense.
- Immune Response: The body activates its defenses, ranging from fever to antibody production, in a concerted effort to reclaim balance, much as a city would rally to clean up after a polluting industry.

Imagine this delicate ecosystem as a communal garden, where the mutualists are like bees that pollinate flowers and also take nectar—both creatures prospering. The commensalists are akin to earthworms enriching the soil, helping the garden thrive without drawing on its resources. In contrast, the parasites are the weeds that might choke out flowers to claim space and nutrients for themselves.

This vivid portrayal brings to light the elegant, though sometimes competitive, nature of microbial relationships in the body, reflecting not just their complexity but their profound significance to our overall health and wellbeing. Through understanding these interactions, one grasives the invaluable role of these microscopic companions in the vast symphony of life within us.

Think of the microbiome as a highly skilled team in the kitchen of a bustling restaurant, where each chef has a specific role that contributes to the creation of a savory meal. In your body, this translates to a group of microorganisms that specialize in breaking down the food you eat into smaller, more manageable parts, much like chefs finely chopping ingredients so they can be better combined and cooked.

These microscopic chefs also have the remarkable ability to whip up essential nutrients from scratch. Take vitamin K, for instance, which is crucial for blood clotting and bone health. The body itself can't make enough of it, but, luckily, these gut microbes can, acting like an in-house vitamin factory. They also lend a hand in synthesizing certain B vitamins vital for energy and brain function.

But their talents don't stop there. Beyond helping with digestion and concocting vital nutrients, they're also like the vigilant security guards of your body, fortifying your immune system. They form barriers against harmful invaders, educate immune cells on potential threats, and even help control inflammation—like bouncers keeping the peace in a crowded venue.

This explanation uncovers just how instrumental the microbiome is in the daily operations of your digestive, nutritional, and immune systems, operating quietly yet impactfully, like the crew behind a grand stage play. By maintaining a healthy microbiome, you're not just supporting these minute workers; you're offering them the best tools to keep the show running smoothly — your health in harmonious balance.

Let's take a closer look at the bustling cityscape of the microbiome, where microorganisms perform tasks akin to workers in a metropolis, each contributing to the health and prosperity of the community within your body.

Digestion is where the magic starts. Think of dietary fibers as unrefined resources trucked into the city. Residents of the gut—beneficial bacteria such as Bacteroides and Lactobacillus—act like industrious factories, fermenting these fibers into short-chain fatty acids (SCFAs). These SCFAs are akin to clean energy for the city, powering up colon cells and forging a strong barrier against diseases, while also providing a critical energy source that furthers the entire metropolis's vitality.

This urban ecosystem produces its own health-boosting commodities, too. Certain gut bacteria double as nutritional chemists, synthesizing essential B vitamins like B12, B7 (biotin), and B9 (folate)—microscopic yet mighty architects of energy metabolism and brain health. One can imagine these vitamins as the city's electricity grid, empowering various neighborhoods— body systems—with the energy to thrive and think.

Further, protective bacteria in the gut resemble the city's peacekeeping force. They patrol the streets—our digestive tracts—to ward off unwanted intruders, being pathogens, through mechanisms of competitive exclusion and direct antagonism. These beneficial microbes can send chemical signals, similar to an alert system, which mobilize the body's own immune cell defenders. They're like liaison officers, briefing immune cells on recognizing enemies should they try to invade again.

The messaging system of this microbe metropolis is incredibly advanced. Immune signals, like cytokines, are dispatched like messages over a city-wide intercom, directing immune cells to quell inflammatory uprisings and restore order within bodily districts when unrest, or inflammation, breaks out.

Envisioning the microbiome in this way weaves a vivid narrative, illustrating each individual's role in this complex yet harmonious arrangement. This web of interdependency illustrates the profound elegance of our inner ecosystem—a place where every microbe has a role, and our well-being hinges on their tiny, collective endeavors to keep the city humming along.

Imagine your microbiome as a charming, meticulously cared-for garden nestled within the walls of your own body. Just as a gardener cultivates the land, balancing sunlight, water, and nutrients to foster a vivid tapestry of blooming flowers, thriving plants, and lush greenery, so must you tend to the diverse ecosystem of your microbiome. A balanced microbiome, like a flourishing garden, is a marvel of diversity and harmony, where different species of bacteria coexist, each contributing to the garden's overall health by breaking down organic matter, deterring pests, or pollinating plants.

In contrast, when this balance is upset, much like an overgrowth of weeds in a neglected garden that can choke out the flowers, harmful microbes can overwhelm the beneficial ones within your gut, disrupting the delicate equilibrium. The result may parallel a garden in disarray, with its beauty wilted and resilience weakened.

Therefore, nurturing the microbiome with thoughtful lifestyle choices, such as consuming prebiotics and probiotics akin to watering and fertilizing the garden, helps maintain this essential balance. By doing so, you're not just upholding a collection of microbial life; you're cultivating a sanctuary that supports your well-being, much like a well-tended garden becomes a bastion of tranquility and abundance. Through this analogy, the complex reality of the microbiome's balance becomes tangible and resonant, reminding us of the simple actions we can take to ensure its and, by extension, our own flourishing health.

Let's take a deeper look at the bustling marketplace of the gut, where prebiotics and probiotics play pivotal roles akin to market gardeners and seasoned vendors. Prebiotics, found abundantly in foods like garlic, onions, and bananas, are comparable to specialized fertilizers that enrich the soil. They provide the kind of nourishment that selectively encourages the growth of friendly bacterial populations, such as Bifidobacteria, much like a discerning gardener who uses premium compost to nourish only the choicest plants.

As these preferred bacteria digest prebiotic fibers, they transform them into powerful substances like short-chain fatty acids (SCFAs). Imagine these SCFAs as the nourishment that fortifies the walls of the intestinal tract, regulates metabolism, and keeps the gut lining robust against unwelcome intruders.

Probiotics, those live and active cultures found in fermented delights like yogurt and kombucha, can be pictured as a caravan of beneficial flora setting up shop in the gastrointestinal tract. They bring with them a wealth of healthy bacteria such as Lactobacillus strains, quickly becoming local 'vendors' that contribute to the gut's diversity and vitality. These newcomers help crowd out harmful microorganisms—potential 'weeds' in the garden—by competing for space and nutrients.

Moreover, these probiotic strains work in tandem with the body's own cells to enhance the gut's barrier function, much like a town's defenses are bolstered by the addition of stout gates and vigilant guards. They also play a diplomatic role, mediating peaceful communication between the immune system and the microbial residents, ensuring the immune system's responses are well-regulated and precisely targeted, thus preventing unnecessary inflammation.

The sumptuous interplay between prebiotics and probiotics within the gut microbiome reveals a tapestry woven with intention and care; it's a delicate arrangement that reflects the wisdom of nurturing a thriving, dynamic ecosystem in your own body. Just as a gardener's choices can lead to a blooming paradise or a faltering plot, one's dietary decisions have the power to craft an inner milieu brimming with life or leave things in disarray. So, with every fiber-rich bite or sip of a fermented brew, imagine the positive bustle you're inspiring in the city that lives within you.

Consider your microbiome as an intricate garden within you, one that thrives on the variety and quality of the nutrients it receives, just as gardens prosper from rich soil and diverse plant life. Your dietary choices are like the seed selections and fertilizers used by a thoughtful gardener. Consuming a rainbow of plant-based foods introduces a wealth of 'seeds' that can blossom into a vibrant floral display of healthy gut bacteria. Lean proteins and complex carbohydrates act as 'slow-release fertilizers,' providing sustained nourishment to your internal ecosystem.

On the other hand, indulging in processed or sugary foods simulates the 'synthetic fertilizers' that might offer immediate, but ultimately unsustainable growth, encouraging invasive 'weeds' - harmful microbes - to flourish.

Meanwhile, antibiotics are like the pesticides of our inner garden; while targeting unwanted pests, their non-discriminatory nature can also sweep away beneficial organisms that contribute to the garden's diversity and defenses. With overuse, just as land might become barren after excessive pesticide application, the microbiome too can suffer a loss of its treasured flora, weakening your body's own bountiful garden.

Through every meal and medication, you wield the power to cultivate this inner terrain. Equipped with knowledge, every food choice becomes a deliberate act of gardening, ensuring the lushness of your internal flora is preserved and enhanced, nurturing not only a garden but the entirety of your wellbeing.

A Gardener's Guide to the Microbiome

Nourishing Your Inner Garden: The Role of Fibers and Probiotics
Imagine your gut as a fertile plot of land where the food you eat is the seed and water that enables the garden to flourish. Fibers from fruits, vegetables, and whole grains act like mulch, fostering growth among

beneficial bacteria. These microbes, in turn, convert fibers into life-sustaining nutrients. Probiotics, found in fermented foods such as yogurt and kefir, can be thought of as the robust, ready-to-bloom plants you introduce to this internal landscape. They integrate seamlessly into your gut garden, enhancing its diversity and resilience against invasive species, the pathogenic bacteria.

Weeding Out the Unwanted: Understanding the Selective Action of Antibiotics

Antibiotics can be likened to a powerful herbicide that targets the harmful, disease-causing weeds but can also inadvertently damage the lush vegetation you've worked hard to cultivate. They don't discriminate, decimating both beneficial and harmful bacteria. Just as a gardener would use herbicides judiciously, so should antibiotics be used sparingly and under the right circumstances, ensuring that the good bacteria in your gut can thrive and keep the 'weeds' at bay.

Cultivating Resilience: Rebuilding a Robust Microbiome Post-Antibiotics

After a course of antibiotics, your gut garden might need rehabilitation to return to its former glory. Consuming prebiotics acts as both fertilizer and soil conditioner, nourishing and supporting the regrowth of healthy microbes. Integrating probiotics into your diet post-antibiotics is akin to replanting your garden with strong, beneficial plants. These steps, combined with a balanced diet rich in plant-based foods, can help your gut microbiome recover and once again bloom with vibrant microbial life.

Remember, a gardener's work is consistent and mindful, with an understanding of each element's role within the ecosystem. Tending to your microbiome with the same philosophy can lead to a vibrant, flourishing internal environment that enhances your overall health and well-being.

A balanced microbiome is like a bustling city where all is in order – the traffic flows well, the parks are green, hospitals operate efficiently, and waste management is effective. In this city, friendly bacteria do the work of keeping the environment – your body – in tip-top shape. They help digest food, fend off harmful germs that might cause diseases, and ensure the immune system operates like a well-trained urban police force, ready to protect residents against health threats.

When this microscopic metropolis is in harmony, the risk of infections, chronic diseases, and even certain allergies is lowered, contributing to the better health of each individual. And just as one person's health can affect others in a community, a population of healthy individuals, each with balanced microbiomes, can have a significant positive impact on public health. Fewer infections mean less transmission, lower healthcare costs, and a focus on prevention rather than cure.

This balance doesn't just prevent illness; it promotes overall health, making recovery from sickness quicker and helping bodies to respond better to treatments when needed. By maintaining this inner equilibrium, a person is not just looking after their own health but is an active player in the well-being of the community at large.

Just as a gardener takes pride in the blooming beauty of a well-kept garden, you too can cultivate a thriving internal ecosystem. It starts with a seed—understanding that every bite taken and every habit formed can nourish or neglect this internal flora. A garden doesn't flourish overnight, nor without effort, and the same goes for your microbiome. It beckons for mindful care: a variety of foods rich in fibers, vegetables reminiscent of a palette of vibrant colors, and fermented foods that add life to the soil of your gut garden.

Yet, the empowerment lies not just in knowing what helps the garden grow but in wielding that knowledge day by day. Making choices that foster this internal vitality isn't a mere health trend; it's an act of nurturing the very essence of life within. With every meal that's more colorful, every decision to steer away from the allure of processed foods, and every thoughtful use of antibiotics as if they were garden tools—to be used with precision and care—you become not just a caretaker, but an artisan of your own well-being.

So let this final thought be a call to action: your microbiome, your exquisite inner garden, awaits your tending. In the cultivation of this garden lies the beauty of robust health and the blossoming of life's potential. Take the trowel in hand, and with each conscious choice, feel the joy of watching your internal ecosystem thrive.

FERMENTATION AND BIOTECHNOLOGY MICROBES AT WORK

Welcome to the tiny yet mighty world of microbes, the unseen forces behind some of humanity's greatest culinary delights and scientific breakthroughs. This narrative unfolds the pivotal role these microorganisms play in the art of fermentation and the rapidly evolving arena of biotechnology. The journey begins with a single yeast cell, invisible to the eye, yet powerful enough to orchestrate the transformation of simple ingredients into complex products like bread, cheese, and even a pint of ale.

Imagine opening the door to a traditional bakery where the air is fragrant with the aroma of fresh bread. This is the handiwork of microbes, tirelessly exhaling gases to make the dough rise. Now, picture a laboratory on the frontier of modern medicine, where biotechnologists harness similar microbes not for their breath but for their ability to produce life-saving drugs. Every step in this exploration is a brushstroke on the broader canvas of biotechnology, painting a world where microscopic organisms work wonders that resonate on a global scale.

Here, technical terms become friends instead of puzzles, and complex processes are broken into stories that unfold with the same ease as flipping through a photo album. One will discover not just the 'how', but the 'why'— why microbes matter, why their work is indispensable, and why understanding them offers a gleaming key to unlocking sustainable innovations. So, with a clear mind and keen curiosity, join in unraveling the stories of these tiny but influential inhabitants of our world as we reveal the magic that brews when microbes are at work.

Fermentation is the alchemy of the food world, a process that might as well be magic if not for the science we understand behind it. Imagine a brewmaster, a guardian of tradition and chemistry, carefully selecting grains and hops, much like a painter chooses the right blend of colors. Just as the brewmaster combines these ingredients, adjusts temperatures, and monitors times to create a perfect batch of beer, so do microbes orchestrate the transformation of simple sugars into the complex flavors found in everything from a crusty loaf of sourdough to tangy kimchi.

These microbial maestros—yeast and bacteria—are like the skilled hands of the brewmaster, diligently working behind the scenes. The hops and malt in brewing are no different from the fresh milk in a vat of soon-to-be cheese or the ripe grapes in a vat of wine, each waiting for microbes to turn them into something more. Yeast, the tiny powerhouse of fermentation, feasts on the sugars, releasing carbon dioxide and alcohol in a bubbly performance that could steal the show at any culinary concert.

This process isn't about just creating food with a punch of flavor—it's about the transformation. Much like turning grapes into a fine wine or barley into an aromatic ale, fermentation elevates basic ingredients to rich and complex delicacies. Each step in the fermentation process is like a stroke of the brewmaster's crafty hand, turning ordinary into extraordinary, showcasing not only the power of these minuscule creatures but also why understanding and appreciating their role opens up a world where food becomes more than sustenance—it becomes a masterpiece of taste and tradition.

Here is the breakdown of the fermentation process—a journey where microscopic creatures brew life's flavorful concoctions within the walls of our foods and beverages:

- **The Beginning - Ingredient Selection:**
  - Just as a brewmaster selects the finest grains and hops, the fermentation process starts with selecting the right 'food' for microbes.
  - Input: Simple carbohydrates like glucose from fruits or maltose from grains serve as the starting ingredients, much like a pantry stocked with essentials.

- **The Main Players - Our Microbial Cast:**
  - **Saccharomyces cerevisiae:** Often referred to as brewer's yeast, this species is responsible for alcohol fermentation, much like a master distiller overseeing the transformation of mash into spirits.
    - Produces: Ethanol and carbon dioxide, creating the bubbly effervescence and alcohol content synonymous with beer and bread rise.
  - **Lactobacillus:** This species is a friendly bacterium, similar to a baker ensuring bread comes out with just the right sourness and texture.

- Produces: Lactic acid, which gives yogurt and sauerkraut their signature sour taste.

- Environmental Conditions - The Brewery Settings:
- Temperature: Just as some beers require colder storage to ferment properly, the microbial fermentation process requires specific temperature settings to thrive and produce the desired flavors.
- pH Levels: The acidity of the environment needs to be just right, like the pH of the soil for the perfect hop growth. Microbes have their preferred pH to effectively transform substrates into products.
- Oxygen Availability: Anaerobic conditions (absence of oxygen) are generally required for brewing alcohol, akin to the closed, pressurized fermenters that keep oxygen out and allow the magic to happen.

- The Process - From Ingredients to Products:
- Fermentation begins when yeast and bacteria consume the sugars, a bit like guests at a feast, breaking down glucose through glycolysis.
- Sublist for **Saccharomyces cerevisiae**:
- Aerobic respiration is the first choice when oxygen is available—this is like the opening act, preparing the stage.
- In anaerobic conditions, alcohol fermentation kicks in—this is the main event, where the simple sugars turn into ethanol and CO_2.
- Sublist for **Lactobacillus**:
- The lactic acid fermentation process is straightforward; it's like a one-dish recipe, turning sugars directly into lactic acid without the need for oxygen.

- The Result - Flavorful Creations:
- The end products of fermentation are as varied as the beer styles in a brewery, with each microbe contributing to a unique taste and texture profile in the food or drink, like signature brews that bear the hallmark of their creators.

Understanding fermentation is to appreciate a molecular gastronomy, where microbes donned with aprons of enzymes and pots of biochemical reactions whip up delights that tantalize our tastebuds. It's not just a culinary journey; it's a heritage, a tradition, and a scientific marvel that reflects the ingenuity of nature and human innovation working together.

Microbes, in their minuscule dimensions, hold the blueprints for an astonishing array of medical breakthroughs. Inside biotechnology labs, the process of coaxing these organisms to produce drugs and vaccines, much like convincing a plant to yield the ripest fruits, unfolds with meticulous precision. Beginning with selection, scientists identify microbial strains that naturally produce beneficial compounds, just as a gardener selects the best seeds. These microbes, such as the mold **Penicillium** that gives us penicillin, become miniature factories, churning out molecules that have the power to heal.

From here, the production scales up. It's comparable to moving from a home kitchen to an industrial kitchen, ensuring that the conditions are optimal for each microbe to work at peak efficiency. In giant vats called bioreactors, microbes undergo a process of fermentation, where they grow and multiply in a carefully controlled environment. Temperature, pH, and oxygen levels are adjusted with surgical precision, as if tuning an instrument before a symphony, to elicit the desired medicinal compounds.

As these microorganisms consume nutrients, they produce biochemical products much like a writer pens words on a page—a careful, deliberate process that results in substances such as insulin for diabetes management or the antitumor drug taxol. Today, the technique of genetic engineering has turned this process into an even more powerful tool, giving us the ability to insert genes coding for specific medicines into microbial DNA—a feat akin to installing a new piece of state-of-the-art machinery in a factory to boost its output.

The significance of these microbial pharmacists in healthcare cannot be overstated. Vaccines, for instance, are often born from the proteins that microbes produce, teaching our immune systems the art of defense as if they are wise sages imparting knowledge. Each medical marvel coming out of microbial production not only shapes the present landscape of treatment but also lights the path toward future wonders.

Understanding these mechanisms instills a deep appreciation for the microbiological savants at the heart of medicine's most potent therapies. It gives insight into a world where the smallest organisms harbor solutions to some of life's biggest challenges, illustrating a story of health and hope that connects the intricacy of microbes to the vast canvas of human well-being.

In the realm of pharmaceuticals, microbes serve as tiny but expert artisans, crafting compounds that become the medicines people rely on every day. The process starts with gene identification, where scientists act as detectives, deciphering the DNA sequences that hold the instructions for making a specific therapeutic protein, just as an artist might select the perfect material for a sculpture.

Next comes the role of plasmids, small circular DNA strands separate from a microbe's chromosomal DNA. They work like a craftsman's tools, carrying these selected genes into microbial cells. Think of these plasmids as a master key, unlocking the microbial machinery to start the production of medicinal compounds.

The process gets intricate as genetic engineering enters the stage. Scientists insert the plasmid toolkits into the microbial workforce—perhaps yeast like **Saccharomyces cerevisiae** or bacteria such as **Escherichia coli**. Once inside, the microbial cell reads the new genetic material as if it's the microbe's own, following its encoded instructions to produce proteins like insulin. The cell's own enzymes, acting as skilled guildworkers, guide the construction of these complex molecules step by step. Temperature, nutrients, and pH are finely tuned, similar to adjusting lighting and pressure to ensure ideal crafting conditions.

From synthesis to harvesting, the microbes culture in bioreactors where they churn out the medicinal product. At the end of this meticulous process, the therapeutic proteins are extracted, purified, and prepared for use, much like polishing a newly forged piece of jewelry to a brilliant shine.

These microbes, once harnessed for brewing and baking, now synthesize vital medical substances. This convergence of biology and technology demonstrates not only a feat of human ingenuity but also the multifaceted capabilities of microbes and the transformative power they have in healthcare. This narrative doesn't just explain a procedure; it brings to light the broader picture of biotechnology's impact on medicine, inviting a deeper respect and understanding of the science that weaves throughout these life-saving advancements.

Think of a field of tall corn swaying in the breeze, not a simple crop but a source of raw energy, much like a pantry full of uncooked ingredients. In the same way a master brewer transforms water, yeast, and grain into an invigorating beer, microbes turn this biomass into biofuel—a renewable energy to power our world. This transformation begins when the biomass, rich with sugars just waiting to be unlocked, enters a bioreactor, the bustling kitchen where microbial chefs get to work.

Here, specialized microbes, trained in the culinary art of fermentation, start to digest the plant materials, breaking them down like a chef meticulously chopping vegetables. Sugars once trapped within the plant fibers are released and gobbled up by these hungry microorganisms. Much as yeast ferments sugars into alcohol in brewing, these microbial maestros ferment biomass sugars into biofuels like ethanol or biodiesel, concocting a potion that can fuel cars rather than people.

The elegance of this process lies in its simplicity and cyclicality. Crops capture solar energy and store it as carbohydrates, just as solar panels soak up sunlight. The microbes then convert this stored energy into fuel, not unlike a generator converting solar power into electricity. It's a process that cycles from sun to fuel tank, a green symphony of nature and technology working in harmony.

Understanding how these microorganisms perform such an intricate ballet, transforming something as earthly as cornstalks into a source of renewable energy, helps one grasp not only the science behind biofuel production but also its importance. By leveraging the natural efficiency of microbes, this process offers a glimpse into a sustainable future, where every stalk and leaf has the potential to keep the engines of human progress running cleanly and efficiently.

Here is the breakdown of the biofuel production process, a marvel of nature's pantry turned into renewable energy much like flour and water are turned into a rising loaf of bread:

- **Biomass Sources Identification:**
    - Like choosing the main ingredient for a signature dish, we select biomass sources such as corn stalks or sugarcane, rich in the sugars needed for fermentation.

- **Pre-Treatment Processes:**
  - Biomass undergoes pre-treatment to unlock its energy potential, akin to marinating meat to tenderize it for cooking. This might involve:
    - Physical processes like milling, to chop the material into smaller pieces.
    - Chemical processes using acids or bases to break down complex carbohydrates into fermentable sugars.

- **Microbial Action in Fermentation:**
  - Microbes such as **Clostridium cellulolyticum** step onto the scene to break down cellulose, turning plant fibers into sugars as a chef transforms ingredients into a stew.
    - These microbes act on pre-treated biomass, much like yeast acts on dough, converting it through fermentation.

- **Key Enzymes in Metabolic Pathways:**
  - Enzymes such as cellulases and amylases serve as the kitchen gadgets, accelerating the breakdown of biomass into simpler forms, much as a blender turns whole fruit into smooth juice.

- **Stages of Fermentation:**
  - Fermentation proceeds in orderly steps, each like a course in a fine meal:
    - Hydrolysis: Where the complex carbohydrates are broken down into simple sugars.
    - Acidogenesis: Where simple sugars are turned into acid by microbes.
    - Acetogenesis: Where these acids are further converted into acetic acid by other bacteria.
    - Methanogenesis: Where acetic acid is transformed into methane by methanogens, completing the process as the final garnish on the dish.

- **Post-Fermentation Processing:**
  - After fermentation, the 'dish' needs refining, akin to straining a broth to achieve a clear consommé.
    - This might involve distillation, where biofuels are separated and purified, getting them ready for use in vehicles or power generation, serving as the well-prepped and sustainable 'meal' for our machines.

Understanding this process, from field to fuel tank, sheds light on a sustainable future where each stalk of biomass contributes to a cleaner, greener world. It's a testament to human ingenuity — using the Earth's natural abundance and the tireless work of microbes to create a renewable source of power that can drive us toward an eco-friendly future.

Microbes play a role in waste degradation and environmental detoxification much like a cleanup crew restores calm after a storm. In nature's toolkit, these microbes are like tiny waste processors, transitioning harmful materials back to harmless earth. A bacterium, for instance, can take a pollutant – something as toxic as an oil spill – and break it down into simpler, safe compounds, similar to how earthworms digest and neutralize waste in the soil.

The process, known as bioremediation, leverages the natural appetite of microbes to consume and transform waste. One can picture these microorganisms as specialized workers, each adept at handling a specific type of rubbish. Some thrive on petroleum, turning slick, contaminating oils into benign substances. Others tackle industrial byproducts like heavy metals, safely sequestering them in a way that parallels a cleanup specialist encapsulating hazardous materials.

This microbial action is central to environmental sustainability. By enlisting these natural decomposers, polluted sites are not just cleansed, but restored to a condition where life can thrive again—much like a desolate lot turned into a verdant park. This synergy between microorganisms and sustainability underscores their pivotal ecological role, providing an organic solution to human-made problems.

Understanding the workings of these tiny environmental stewards offers a vivid portrait of the circular nature of life. It instills a sense of respect for how microorganisms, often out of sight, continuously uphold the delicate balance of our ecosystems. Their silent work is a cornerstone of ecological health and a powerful ally in human efforts to maintain a livable planet.

Let's take a deeper look at the microscopic world of bioremediation, where hidden beneath our feet and in our waters, a tireless legion of microbes is hard at work cleaning up our environmental messes as if they're nature's own little janitors.

- **Common Pollutants and Microbial Detoxifiers:**
  - Hydrocarbons, found in oil spills, often meet their match in the likes of the **Pseudomonas** and **Alcanivorax** species, known for their oil-guzzling appetite.
  - Heavy metals like lead and cadmium are deftly managed by bacteria such as **Cupriavidus metallidurans**, which can transform these toxic elements into less harmful forms, akin to a blacksmith refining raw ore into valuable metal.
  - Pesticides, the persistent scourges of soil and waterways, are broken down by specialized fungi and bacteria, including strains of **Trichoderma** that act like eco-friendly disposal units.

- **Biochemical Pathways for Pollutant Breakdown:**
  - Taking hydrocarbons as an example, enzymes such as oxygenases play the pivotal role of catalysts that would make any chef envious, chopping molecular bonds with the precision of a gourmet knife, making contaminants more digestible to microbes.

- **Influential Environmental Factors:**
  - The temperature, pH, and levels of oxygen available can dictate how effectively our microscopic cleaners work. Warm conditions may encourage them, just as a sunny day might inspire a flourish of garden growth, while too much acidity in their surroundings can be as troublesome as a kitchen filled with smoke, stifling their efforts.

- **Real-World Environmental Success Stories:**
  - A shining example is the use of **Pseudomonas** species in remediating oil spills, where they descend upon the hydrocarbons like a swarm of hungry locusts to a field, leaving behind cleaner waters, mirroring a field ready for new growth.

Understanding these intricate processes reveals the scale and scope of bioremediation's potential beautifully. It's like being aware of the bustling life in a hidden city whose inhabitants toil to keep it running smoothly. By delving into the specifics—by pairing pollutants with their microbial matches—we paint a detailed picture of the sophisticated interplay between biology and the well-being of our planet. This is a testament to the power of life, no matter how small, to restore balance and order in the world.

Picture this: in the rolling hills of France, the ancient art of cheesemaking takes on a new science fiction twist, where microbial biotechnology steps in like a modern-day alchemist. It's here that the humble **Penicillium roqueforti** fungus plays its critical role in creating the bold veins of Roquefort cheese, turning milk into a culinary treasure as if by magic. This little organism, when added to cheese, works its way into the heart of the dairy, crafting the sharp, tangy flavors much like a master chef inventing a new, sought-after recipe.

Not all microbes, though, are busy whipping up gourmet delights. Some are out in the field, making a difference in less appetizing, yet equally important ways. Take the Exxon Valdez oil spill for example, a true environmental calamity where **Pseudomonas** bacteria became the unlikely heroes. These microbes, engineered to thrive on crude oil, descended upon polluted shores like an army of tiny cleaners, breaking down the spilled oil into less harmful substances in a process as efficient as nature's own recycling system.

These stories span from the delightful to the crucial, highlighting how microbial biotechnology sneaks into various facets of life. They remind one that sometimes, it's the smallest beings that have the grandest impact, shifting the course of industries and aiding in the healing touch nature sometimes requires. It's here, in these practical examples, that the boundless potential of microbial biotech is wholeheartedly felt, fostering a delight for both the cheese aficionado and the eco-warrior alike, and leaving us with a profound appreciation for the microcosms that sculpt our everyday experiences.

Let's take a closer look at the roles that specific microbes play in the realms of gastronomy and environmental corrections, starting with a microbial protagonist in a tale of cheese-making. Imagine **Penicillium roqueforti** as an expert artisan, carefully weaving blue veins of flavor through the canvas of curd. This mold begins its life as a spore, which, when introduced to the curd, germinates and sends out threads of mycelium, much like roots in a flourishing garden. As it grows, it releases enzymes that break down fats and proteins, slowly painting the bold, tangy flavors we savor in Roquefort cheese, turning the curd into a masterpiece of edible art.

Now, shifting focus to the aftermath of an environmental disaster, imagine **Pseudomonas** species as nature's own cleanup crew, arriving at the scene of a spill. These bacteria harbor enzymes like oxygenases, which act like tiny molecular scissors, snipping the long chains of hydrocarbons in petroleum into smaller, more manageable fragments. Each cut by these enzymes detoxifies the oil, transforming it into substances that blend harmlessly with the environment, turning a tragedy into a triumph of restoration.

There might also be a twist in the plot, where scientists draft in genetic engineering to give these microbes an upgrade. By introducing new genes into their DNA, the abilities of these tiny workers are amplified, enhancing their efficiency at their tasks. Whether that means crafting more delicious cheeses or cleaning up spills faster and more thoroughly, genetic engineering stands as a powerful director, orchestrating the microbial performance for the greater good.

In these stories, microbes take on the role of unsung heroes, small in size but immense in impact. It's a delicate dance they perform, guided by their genetic blueprint and the environment they inhabit. Understanding these processes, the viewer comes to appreciate the complex interplay of life, science, and the outcomes we depend on, seeing not just isolated acts, but a continuous journey towards innovation and recovery.

In closing this chapter, let's imagine how the ongoing quest to harness, enhance, and expand microbial biotechnology might transform landscapes of industries the world over. Picture a future where oil spills are rendered powerless by swarms of oil-eating microbes, leading to a sharp decline in marine disasters. Envision fabrics woven with color by eco-friendly bacterial dyes, reducing the reliance on harmful chemicals. Visualize farms where natural microbial pesticides protect crops without the environmental toll of synthetic alternatives.

The nexus of nature and technology offers an avenue to solutions once thought impossible—a world where waste becomes a valuable resource, and once-contaminated lands bloom anew, much like the transformation of barren fields into fertile farmland. The role of microbes in these advancements, akin to the masterful strokes of an artist, shapes a sustainable dawn of industrial evolution.

As new chapters of microbial biotechnology unfold, it's the meticulous study of these organisms that will pave the way for their orchestrated application. From the pharmaceutical labs crafting next-generation antibiotics to the environmental cleanup sites turning pollutants into energy, every application speaks to the versatility and ingenuity behind microbial processes.

While delving into the story of these microscopic powerhouses, one is reminded of the grandeur hidden in the minuscule, and the monumental promise it holds. It's a narrative that interlaces the threads of ecological mindfulness with the fabric of human innovation, crafting a tapestry of possibilities that are as wide as the collective imagination. This narrative doesn't simply conclude; it beckons one forward, inviting minds to explore the potential and propel these eco-friendly solutions from visionary ideals into pragmatic, real-world applications.

OUTSMARTING MICROBES THE FIGHT AGAINST DISEASE

Our tale begins in the annals of ancient history, where humanity's understanding of diseases was shrouded in mystery, much like early explorers gazing upon uncharted lands. Through the ages, as symptoms ravaged communities, the source of these ailments remained invisible to the naked eye, and thus treatment was as elusive as a whispered legend. The turning point came with the realization that microorganisms—bacteria and viruses—are the culprits behind many diseases, revealing themselves like hidden enemies in a great war.

The breakthroughs were monumental, comparable to the discovery of new continents in our exploration of health. The accidental unearthing of penicillin by Alexander Fleming marked a watershed moment, offering humanity its first effective weapon against bacterial infections. It was as if a locksmith had crafted a master key capable of thwarting a notorious burglar. This discovery was not just a remedy but a beacon of hope, igniting a pursuit of knowledge and empowering scientists like never before.

This narrative follows the milestones of medical innovation, showcasing how every subsequent discovery, such as the polio vaccine or advanced antivirals, built upon the last. Each advancement is a piece of a larger puzzle, fitting together to form a shield that protects the vast expanse of human health. These achievements in outsmarting microbes weren't without their challenges, however. Just as a master key might both unlock doors yet fail against new locks, resistance and the emergence of new pathogens tested the resilience and ingenuity of science.

Throughout this exploration into the battle against diseases, a clear picture emerges: the fight is as dynamic as it is enduring. Every challenge encountered and overcome serves as a testament to the human spirit and the unyielding quest for survival. As the reader is guided through the intricacies of the microbial world and the defenses erected against it, the narrative encourages an appreciation for the equilibrium between human progress and respect for the unseen natural forces that surround us. This is a journey of how, armed with knowledge and innovation, one navigates a path through uncertainty to forge a healthier future for all.

Microbes are minuscule organisms, invisible settlers on every surface, every drop of water, and every patch of earth. They include a wide array of beings, from bacteria and viruses to fungi and protozoa, each with its unique role in the tapestry of life. Like individual characters in an expansive novel, some microbes can be protagonists, promoting health by aiding digestion or contributing to the earth's nutrient cycles. Others act as antagonists, responsible for infections and illnesses that disrupt this harmonious balance.

For example, consider the bacteria in yogurt, Lactobacillus, fermenting milk into a creamy food enjoyed worldwide. These microbes convert lactose into lactic acid, which gives yogurt its tangy flavor, acting much like skilled chefs perfecting a recipe. Conversely, there are the notorious pathogens, like the Mycobacterium tuberculosis bacterium, which stealthily invades lung tissue, leading to tuberculosis. This harmful interaction resembles a lock-picking burglar quietly breaking into a home to wreak havoc within.

Microbes interact with their surroundings incessantly, communicating with one another and other organisms, exchanging genetic material, and transforming their environments – all on a scale equivalent to a bustling city's economy in just a teaspoon of soil. Some of these interactions are beneficial, such as nitrogen-fixing bacteria replenishing soil fertility like gardeners nurturing plants. Others can be harmful, resulting in sickness and spoiling food, mirroring vandals that disrupt the daily flow of life.

In outlining the roles microbes play, it becomes evident that the relationship between these microorganisms and the larger world is as complex as it is crucial. This chapter will unravel the tiny threads that weave the vast fabric of biological interactions, each microbe a vital stitch holding the material of ecosystems and human health together. Understanding microbes leads one to recognize the intricate connections that make up the web of life, revealing the delicate balance that science works tirelessly to maintain and restore.

Let's take a deeper look at the intricate dance of biochemistry that allows microorganisms to play such pivotal roles in transforming their environment and affecting human health. Imagine Lactobacillus bacteria in yogurt as master artisans, each one working tirelessly to sculpt milk into this beloved food. They do this by fermenting lactose, milk's sugar, into lactic acid through a sequence of steps akin to a carefully choreographed dance. Enzymes such as lactase initiate the process, splitting lactose into glucose and galactose, which are then further processed by another set of enzymes, eventually producing the tangy lactic acid that is the hallmark of yogurt.

In the microbial world, knowing a fellow microbe's tricks is invaluable, akin to chefs swapping secret recipes. They exchange genetic material through processes named transformation, where they pick up stray DNA snippets from their environment; transduction, akin to a delivery service, where viruses accidentally transfer genes between bacteria; and conjugation, which resembles a firm handshake, allowing a direct transfer of DNA from one bacteria to another. This microbial mingling boosts their toolbox of survival skills and adaptability.

Now picture the bacteria Mycobacterium tuberculosis like a crafty invader, quietly infiltrating the lungs. It does more than just enter; it cloaks itself in a suit of armor to evade the body's sentinels, the immune system. Through stealthy maneuvers and hijacking the host's cellular machinery, it creates a stronghold known as a tubercle. Inside, the bacteria thrive, multiply, and eventually spread, leading to the devastating symptoms of tuberculosis.

Understanding these biochemical processes demystifies the unseen world of microorganisms. It shines a light on the delicate interplay of forces within our bodies and the ingenuity of these tiny beings that can both sustain and challenge life. It's a narrative that compels not only respect for the microcosm within and around us but also an appreciation of the scientific endeavors that aim to harness these processes for the advancement of human health.

Imagine the medical field as a sprawling battlefield, one where humanity had long been on the defensive against an invisible yet formidable foe: bacterial infections. The discovery of penicillin was like the arrival of an unexpected ally, swinging the momentum in favor of the home team. It wasn't just a new weapon; it was a game-changer, akin to the underdog finding a secret passage that could outflank a seemingly unbeatable adversary.

Before penicillin, even a minor scrape could lead to life-threatening infections. Doctors, much like generals without a crucial part of their arsenal, were often powerless to stop the bacterial onslaught. But penicillin, the product of Alexander Fleming's serendipitous discovery, arrived like the revelation of a powerful strategy forgotten through the ages, suddenly remembered in the eleventh hour of conflict.

This antibiotic became the first of a mighty medicinal phalanx, thriving in the treatment of infections that were once thought invincible. Like a mythical sword that can cut through the toughest armor, penicillin cleaved the defenses of bacterial cells, leading to a cascade of victories across the world. Its effect on medicine wasn't subtle; it was as dramatic and transformative as the invention of the printing press on the spread of knowledge—a beacon of hope that turned dark times into an era of recovery and advancement.

By making such an immense leap forward, penicillin set the stage for an age where infections could be controlled and conquered, saving countless lives. It illustrated the power of a single scientific breakthrough to alter the course of history, elevating humanity in its continuous quest for healing and wellbeing.

Here is the breakdown of how penicillin, a molecule no bigger than a crumb, stands as a colossus against bacterial infections:

- **The Structure of Penicillin:**
    - Picture penicillin as a bespoke key, expertly designed to match the locks on the doors of bacteria cell walls.
    - Its core, the beta-lactam ring, is the part of the key that fits precisely into the bacterial enzymes known as penicillin-binding proteins (PBPs).
    - This exact fit prevents the PBPs from synthesizing peptidoglycan, the main material of bacterial cell walls, similar to a locksmith tool that can prevent a lock from being constructed.

- **Stages of Bacterial Cell Wall Disruption:**
    - Initiation: Penicillin attaches to PBPs, much like a key sliding into a lock.
    - Inhibition: The beta-lactam ring disrupts the PBP's action, halting the construction of the peptidoglycan mesh, like a jammed lock halting the opening of a door.
    - Cascade Effect: With the cell wall synthesis interrupted, the bacteria are unable to maintain their rigid structure. This leads to swelling and bursting

of the cell, much like a water balloon bursting without its intact skin.

- **Bacterial Resistance Development:**
 - Mutation and Adaptation: Bacteria develop resistance through mutations, equipping themselves with new types of locks or even lock-pickers like beta-lactamases, which can break the beta-lactam ring of penicillin.
 - Bacteria can acquire these mutations naturally or receive them from other bacteria, almost as if the locks had an emergency meeting and changed their design in response to a breach.
 - Human Response: In retaliation, scientists develop new generations of antibiotics, akin to locksmiths crafting sophisticated keys to counter the newly developed locks.

By understanding penicillin's strategic infiltration into the bacterial cell's architecture, one gains a deeper appreciation for one of the most significant skirmishes in medical history. It's not just a tale of scientific triumph but a reminder that, in the ever-evolving battlefield of antibiotics and bacterial resistance, the ingenuity and persistence of human innovation continue to play a critical role.

To unravel the mystery of how vaccines act as the body's trainers, preparing it to fend off diseases, start by picturing the body as a fortress. The immune system is the fortress's sentry, ever-vigilant against invaders like viruses and bacteria. Now, a vaccine enters the scene like a set of blueprints, revealing the vulnerabilities of a potential invader to the sentry without exposing the fortress to danger.

Inside the vaccine are tiny pieces of the invader, like a most-wanted poster with just enough details to recognize the culprit. These bits are harmless, but they are the key to training. When a vaccine is administered, the immune system's cells scrutinize these pieces and remember their features. This is akin to the sentry making meticulous notes about a potential intruder's profile.

Following this training, if the real enemy ever dares to attack, the immune system recalls the intruder's profile from the vaccine's wanted poster. It then mounts a swift, targeted response, deploying antibodies—like sentry's arrows—to neutralize the invaders. The beauty of vaccination lies in this rehearsal; it preps the body's defenses for a real encounter without the risk of an actual battle.

Breaking down this process reveals the intelligence behind vaccines—none of this would be possible without the body's remarkable ability to learn from and adapt to these mock drills. The result is a shielded and prepared body, a fortress less likely to succumb to the invader should it attempt a siege. Vaccination is, therefore, not just a medical marvel; it's an orchestrated rehearsal for a fight that plays out within, guarding health and preserving well-being.

When a vaccine enters the battlefield of the body, it kick-starts a series of strategic events, marking the commencement of a critical drill for the immune system. Here's how it unfolds, step by decisive step:

First up is the initial recognition phase. Consider antigen-presenting cells (APCs) like scouts on a reconnaissance mission. They encounter the vaccine's components, which are essentially harmless look-alikes of the pathogen, called antigens. The APCs consume these antigens and present fragments of them on their surfaces, essentially placing a 'Wanted' poster for the immune system to see. This alerts the body: 'Pay attention; this is the enemy.'

Next is the lymphocyte activation phase. It's time to muster the troops. The displayed fragments catch the attention of T-helper cells—a type of lymphocyte that acts as the commanders in the immune response. They analyze the information and send out orders, enlisting B-cells, which are like the body's specialized infantry ready to produce targeted weapons against the pathogen.

Then comes the antibody production phase. Picture B-cells, now with the plans to the enemy's castle, rapidly multiplying and morphing into plasma cells, which are akin to the blacksmiths of the immune system. These plasma cells forge vast amounts of antibodies—proteins tailored to neutralize the pathogen, much like crafting swords designed to fit perfectly into the chinks of an invader's armor.

Finally, the memory formation phase ensures this isn't a one-off battle but a legacy of defense. Some B-cells become memory cells, acting as the fortress's chroniclers. They record everything about this invader so that if it dares attack in the future, they will be ready to prompt a faster and stronger immune response—like an old battle plan pulled from the archives and used to outmaneuver a returning adversary.

By journeying through this microscopic choreography, one gains a profound respect for the immune system's capability to learn and adapt, showcasing vaccination as a preemptive strike that teaches the body how to protect itself without facing a real threat. It's a testament to how understanding and technology converge to outsmart nature's challenges and safeguard health.

As if ripped from the pages of a mystery novel, antibiotic resistance emerges as the crafty villain in healthcare's ongoing narrative, a master of disguise cloaked in a shroud of enigma. It is as though every time the heroic doctors think they have the upper hand, this adversary morphs, presenting a fresh set of riddles wrapped in a puzzle.

Imagine a lock that changes its shape just as you think you've crafted the perfect key. That's the game of deception and evolution antibiotics face with resistant bacteria. With each course of antibiotics, it's conceivable to wipe out susceptible bacteria, much like a team of detectives catching the usual suspects. But lurking in the shadows are the few who've learned to elude capture, passing on their secrets like coded messages among spies.

These resistant strains are the bacteria that have learned to dodge the antibiotic's fatal blow, much like a virus in a computer system that becomes immune to the standard anti-virus software. They force medical researchers back to the drawing board, prompting a constant flux of innovation akin to detectives needing ever more sophisticated tools to catch an increasingly evasive criminal.

This ongoing battle, however, isn't just a captivating scientific duel; it underpins the very efficacy of modern medicine. The rise and spread of these superbugs question the sustainability of current treatment approaches and remind us of the relentless adaptability of nature. Engaging with this concept not only educates but highlights the real-world implications of antibiotic misuse and the importance of continued vigilance and innovation in the medical community. It's an epic story of adaptation and survival, one where every advance in the quest for health opens a new chapter in the saga of resistance.

Let's take a deeper look at the cunning escape artists of the microbial world: antibiotic-resistant bacteria. In an espionage-filled tale of survival, bacteria have developed several tricks to dodge the once-fatal blows of antibiotics.

- **Molecular Alterations:**
- Think of the bacterial cell as a secure vault, and the antibiotic as a uniquely designed key meant to unlock and neutralize it. Now, imagine the bacterium changing its lock's internal mechanism—a genetic mutation that happens almost as if the lock itself has grown wiser to the attempts of entry. These mutations can affect the antibiotic's target site directly, making the antibiotic key an ill fit and unable to do its job.

- **Survival Strategies:**
- Bacteria have amassed a toolkit of strategies to survive. One such tactic is to produce enzymes, fighters in their security detail, that deactivate the antibiotic. Picture these enzymes as sophisticated gadgets that a spy might use to jam signals or dissolve ropes: they break down the antibiotic's structure, rendering it harmless to the bacterial cells.

- **Horizontal Gene Transfer:**
- In the realm of bacteria, secrets are shared freely, not just through generations, as one might pass down a family recipe, but among peers and even across species. Through mechanisms like conjugation (imagine a handshake that transfers crucial documents), transformation (picking up loose intel from the environment), and transduction (viruses acting as unwitting couriers), resistance traits spread from one bacterial cell to another, arming a larger cohort with the means to resist antibiotics.

This intricate web of defense and counter-defense paints a living tableau of the warfare raging in the microscopic world, with repercussions that scale up to global health concerns. Grasping the many ways bacteria learn to withstand antibiotic onslaught is, therefore, paramount—much like understanding a rival's strategies in a tense game of chess. It elucidates the need for prudent use of antibiotics and for fostering continued innovation in pharmaceutical defenses, reminding us that in this microscopic espionage thriller, the stakes are nothing less than the future of effective medicine.

When new diseases and elusive pathogens break onto the scene, modern medicine responds with a sense of urgency that rivals the most decisive emergency protocols. Picture a quiet night suddenly shattered by the clamor of alarm bells, as first responders suit up, ready to combat an unforeseen blaze. In the world of health, the alarm is the detection of a novel threat, one that carries the potential to upend lives and ripple through communities.

The emergence of these threats triggers a chain reaction akin to a well-orchestrated evacuation. Scientists and physicians scramble like a specialized response team, gathering intelligence on the behavior, weaknesses, and impact of the new pathogen. Laboratories worldwide become hives of activity, buzzing with the urgency to understand and construct a defensive strategy. This entails identifying the pathogen's genetic blueprint, understanding how it invades the body, the symptoms it causes, and how it spreads—all steps as crucial as understanding the source of a fire to effectively douse it.

Each development in this process, from initial recognition to the deployment of countermeasures like vaccines or treatments, must be executed with precision and speed, for the cost of delay can be grave. It's a high-stakes race against an opponent that knows no boundaries and respects no rules, a race where every second shaved off the response time can equate to lives saved.

Detailing this process, one sees the true scope of modern medicine's capability to adapt and react, channeling collective expertise and resources for a common goal: to protect and preserve health. It unveils a testament to human resilience and ingenuity, transforming what once may have been a death sentence into a challenge that can be met, managed, and, ultimately, overcome.

Here is the breakdown of the orchestrated response that modern medicine mounts in the face of new diseases:

- **Detection and Identification Phase:**
    - Initial Outbreak: Comparable to an inspector arriving at a crime scene, epidemiologists observe an unusual spike in illnesses.
    - Lab Analysis: Like detectives in a forensic lab, scientists conduct tests

to identify the pathogen, using tools such as PCR to decipher its genetic code.
- Pathogen Classification: Building a profile of the perpetrator, this involves determining the family and species of the pathogen, much like categorizing a criminal.

- **Research and Development Phase:**
- Understanding Infection: Scientists investigate how the pathogen invades cells, similar to security experts studying surveillance footage to understand a thief's methods.
- Drug and Vaccine Development: In this quest, compounds are tested as if trying different keys to unlock a specially designed safe - searching for one that will effectively neutralize the pathogen.

- **Clinical Trial and Regulatory Approval Phase:**
- Safety and Efficacy Trials: Clinical trials are to medicine what pilot simulations are to aviation - rigorous tests to ensure performance under real-world conditions.
- Regulatory Review: Health authorities carefully review trial data, akin to a legal team examining evidence before a case goes to trial.

- **Manufacturing and Distribution Phase:**
- Production Scale-up: Once an intervention is approved, production ramps up, much like a factory shifting into high gear to meet the demand for a new product.
- Distribution Logistics: Vaccines and treatments are shipped worldwide, an operation as complex as coordinating an international relief effort after a natural disaster.

- **Public Health Implementation Phase:**
- Vaccination Campaigns: Mass vaccination efforts unfold with the precision of a military campaign, targeting vulnerabilities in the disease's spread.
- Monitoring and Response: Health officials monitor the treatment's impact, adjusting strategies as needed, like generals on a battlefield directing their forces to counter shifting threats.

Through this meticulous progression from alarm to action, a panoramic view of the medical community's defense strategy against emerging pathogens unfolds. It's a narrative that showcases not only the scientific rigor and logistical might at humanity's disposal but also the collective determination to safeguard health on a global scale. Each step is a testament to the unwavering commitment to turn scientific insight into life-saving action.

Imagine a detective from the pages of a noir novel, always a step ahead thanks to an arsenal of gadgets; that's the embodiment of modern biotechnological strategies in thwarting diseases. Picture these as the detective's night-vision goggles, seeing through the darkness of our biological unknowns, or as communicators that can intercept microbial messages, foiling the plots of infectious agents before they unfold.

In the world of disease prevention and treatment, these gadgets are the breakthrough therapies derived from the manipulation of genetic material, the designing of bespoke drugs that fit a pathogen's vulnerabilities like a lock and key, and vaccines that train the body with lifelike simulations of a bacterial or viral breach.

This biotechnological toolkit does not stop with the equipment we have today; it evolves as quickly as the diseases it combats. It ranges from CRISPR technology, a precise gene-editing tool reminiscent of a detective fixing an incorrect record in a database, to mRNA vaccines, which, like an informant, give our immune systems the insider knowledge needed to stop a disease in its tracks before it ever poses a real threat.

The promise of these advancements in biotech is the edge they provide, keeping us one step ahead in the perpetual chase of health and wellbeing. They are reshaping the landscape of medicine, offering unprecedented precision and adaptability in the face of ever-changing adversaries. Like an ever-growing detective's toolkit, biotechnology equips us with more ways to outsmart those wishing to do us harm, ensuring that we can look to the future not with apprehension, but with assured confidence.

Let's take a deeper look at CRISPR and mRNA vaccine technology, two of the most heralded protagonists in the narrative of modern medicine.

CRISPR, short for Clustered Regularly Interspaced Short Palindromic Repeats, operates like the meticulous hands of a master watchmaker. Here's how it unfolds:

- Scientists first design a piece of RNA with the exact sequence of a target gene, much like a blueprint for the watch's gears.
- This RNA guide is paired with an enzyme, Cas9, which acts like the watchmaker's tools, precise enough to make even the tiniest adjustments.
- Together, they navigate to the DNA strand in a cell, and the RNA guide ensures they arrive at the exact location of the gene to be edited.
- Cas9 then snips the DNA, removing the faulty segment like a malfunctioning gear, and depending on the desired outcome, the cell either repairs the snip, disabling the gene, or a new segment of DNA is inserted, like placing a shiny, new cog in the watch mechanism.

Switching gears to mRNA vaccines, these function similarly to a cybersecurity team protecting a network:

- Scientists first sequence the genome of a pathogen to identify the blueprint for a key protein—its antigen.
- They then synthesize messenger RNA (mRNA) that encodes the instructions for this antigen, akin to a cybersecurity software update designed to combat a new threat.
- Once inside the body, our cells read the mRNA instructions and produce the antigen, much like a network running a simulation of an attack to expose vulnerabilities.
- The immune system then learns to recognize and neutralize the invader without the danger of a full-blown infection, preparing it for any real future attacks, much as a cybersecurity drill prepares networks for potential breaches.

And amidst these technological advances, there is a sensitive balance—akin to the ethical debate on surveillance—that must be struck with ethical considerations and challenges:

- CRISPR's capability to alter DNA brings forth questions about the implications of gene editing and the potential for unintended consequences, just as precise tools can sometimes adjust the wrong component if not used with caution.
- Similarly, the speed and novelty of mRNA vaccines necessitate rigorous long-term studies to ensure their persistent efficacy and safety, parallel to the ongoing revisions needed to maintain robust cyber defenses.

Both CRISPR and mRNA vaccine technologies represent pinnacle achievements in biotechnology, but with great power comes great responsibility. It is crucial to proceed with care, transparency, and ethical oversight, ensuring that these innovations truly serve the cause of advancing human health and not just the pursuit of scientific possibility.

As this chapter closes, one is left standing at the precipice of an inspiring vista, peering into the future of human triumph over microbial foes. The tale thus far—a saga of intellect and ingenuity—mirrors the finale of a classic detective story, where the solutions to a confounding set of challenges are neatly laid before the reader, promising a sequel of even more inventive and bold stratagems lying just beyond the horizon.

Just as an astute detective pieces together the trail of clues to outwit a cunning adversary, scientific minds across the globe piece together the puzzles of biology, one breakthrough at a time. It is as if each discovery is akin to finding a new key for a series of increasingly complex locks, opening up possibilities that once dwelled solely in the realm of imagination.

This journey through the microscopic battleground has showcased the deft maneuvers and strategic innovations that underpin medicine's responses—a testament not only to human knowledge but to its resilience. These are the stories that are written in the laboratories and clinics, on the front lines of public health, and deep within the cells of our bodies.

So let the final words of this chapter reflect not an end but a continuation, a prologue to the unwritten adventures that await. It's the promise of a future where humanity's potential to innovate shines bright against the shadows cast by diseases, emboldening the spirit to look forward to the next chapter with confidence and eager anticipation for the innovations that tomorrow may bring.

CONCLUSION

As we close the final pages of 'Microbiology Made Easy', it is my hope that the microscopic world has unraveled before your eyes, transitioning from a mosaic of complexities to a vivid narrative of life on the smallest scales. We began this journey by casting light on the minuscule denizens of our world, using the power of analogy to breach the walls of abstraction that often govern our understanding of microbes.

Reflecting on our journey, it becomes clear that the threads of microbiology are interwoven with the very fabric of our existence. We've elucidated the roles of bacteria, viruses, fungi, and protozoa, not as distant, mysterious entities, but as familiar characters playing their parts in the grand production of life. From the roles of microscopic organisms in health and disease to their impact on the environment and their central place in biotechnological innovation, the lessons have been as abundant as they have been enlightening.

This book aimed to demystify advanced concepts and to instill an appreciation for the subtle yet profound influence of these tiny organisms. The significance of microbial cooperation, competition, and adaptation has more than academic value—it touches on the very nature of our survival and prosperity in a rapidly evolving world.

Through engaging analogies and vivid examples, we've pieced together how molecular weapons fight disease, how genetic information flows within and between cells, and how life adapts to challenges on the smallest of stages—a tableau that speaks to the ingenuity and resilience of life itself.

As we reach the end, it is my hope that 'Microbiology Made Easy' serves not merely as a reference but as a lens through which to view the world. May it foster a sustained curiosity and an enduring respect for the unseen forces that shape our lives. May the knowledge imparted here be a bastion against the spread of misinformation and a beacon to guide responsible stewardship of our biological treasures.

So, as you step back into the world, I encourage you to carry with you the understanding that every breath, every morsel of food, every unthought-of moment is a testament to the silent yet bustling activity of microbes. Their saga continues, and so does our exploration. May you look upon this invisible world not as an alien other, but as a wondrous extension of our own—a sphere where the smallest entities conduct the mightiest orchestras. It is my pleasure to have been your guide on this microscopic odyssey.

ABOUT THE AUTHOR

Jon Adams is a Prompt Engineer for Green Mountain Computing specializing and focusing on helping businesses to become more efficient within their own processes and pro-active automation.

Jon@GreenMountainComputing.com

Check out Jon Adams other books of the Made Easy series here:

https://www.amazon.com/gp/product/B0CZHJV6J6?ref_=dbs_p_pwh_rwt_anx_a_lnk&storeType=ebooks

If you enjoyed this book, please leave a review if you have the time!

Printed in Dunstable, United Kingdom